OUT OF OUR HEADS

OUT OF OUR HEADS

Why You Are Not Your Brain,

and Other Lessons from

the Biology of Consciousness

ALVA NOË

🔲 Hill and Wang

A division of Farrar, Straus and Giroux

New York

Hill and Wang
A division of Farrar, Straus and Giroux
18 West 18th Street, New York 10011

Grateful acknowledgment is made for permission to reprint the following
material: Excerpt from Francisco Varela's translation of Antonio Machado's
poem "Laying Down a Path in Walking," originally published in *Gaia, A
Way of Knowing: Political Implications of the New Biology*, edited by
William Irwin Thompson (Hudson, NY: Lindisfarne Press, 1987), pp. 48–64.
Reprinted by permission of Amy Cohen Varela and William Irwin Thompson.

The Library of Congress has cataloged the hardcover edition as follows:
Noë, Alva.
    Out of our heads : why you are not your brain, and other lessons from the
biology of consciousness / Alva Noë.
        p. cm.
    Includes bibliographical references and index.
    ISBN: 978-0-8090-7465-5 (hardcover : alk. paper)
    1. Consciousness.   I. Title.

QP411.N599 2009
612.8'233—dc22

2008031399

Paperback ISBN: 978-0-8090-1648-8

Designed by Jonathan D. Lippincott

www.fsgbooks.com

1   3   5   7   9   10   8   6   4   2

To the memory of Susan L. Hurley

I no more wrote than read that book which is
the self I am . . .

—Delmore Schwartz

CONTENTS

PREFACE

We live in a time of growing excitement about the brain. Only the preoccupation with finding the gene for everything rivals today's widespread optimism regarding all things neuroscientific. Perception, memory, our likes and dislikes, intelligence, morality, whatever—the brain is supposed to be the organ responsible for all of it. It is widely believed that even consciousness, that Holy Grail of science and philosophy, will soon be given a neural explanation. In this era of expensive and flashy new brain-imaging technologies (such as functional magnetic resonance imaging and positron emission tomography), hardly a day goes by without the science pages of our leading newspapers and magazines publishing reports of important breakthroughs and new discoveries.

After decades of concerted effort on the part of neuroscientists, psychologists, and philosophers, only one proposition about how the brain makes us conscious—how it gives rise to sensation, feeling, subjectivity—has emerged unchallenged: we don't have a clue. Even enthusiasts for the new neuroscience of consciousness admit that at present no one has any plausible explanation as to how experience—the feeling of the redness of red!—arises from the action of the brain. Despite all the technology and the animal experimentation, we are no closer now to grasping the neural basis of experience than we were a hundred years ago. Currently, we lack even a back-of-the-envelope theory about what the behavior of individual cells contributes to consciousness. This in itself is no scandal. It is a scandal if we allow the hype to obscure the fact that we are in the dark.

It is sometimes said that the neuroscience of consciousness is in its infancy. But that's not quite right, as it suggests that progress will take care of itself: it's just a matter of time and the normal process of maturation. A better image might be that of inexperienced hikers out on the trails without any clear idea where they are: they are lost and don't even know it! I am writing this book to help us figure out where we are and to show us the way forward.

In a way our problem is that we have been looking for consciousness where it isn't. We should look for it where it is. Consciousness is not something that happens inside us. It is something we do or make. Better: it is something we achieve. Consciousness is more like dancing than it is like digestion.

The aim of this book is to convince you of this. I also want to show you that this is what a genuinely biological aproach to the study of mind and human nature teaches us. The idea that the only genuinely scientific study of consciousness would be one that identifies consciousness with events in the nervous system is a bit of outdated reductionism. It is comparable to the idea that depression is a brain disease. In one sense, that is obviously true. There are neural signatures of depression. Direct action on the brain—in the form of drug therapy—can influence depression. But in another sense, it is obviously not true. It is simply impossible to understand why people get depressed—or why this individual here and now is depressed—in neural terms alone. Depression happens to living people with real life histories facing real life events, and it happens not only against the background of these individual histories but also against the background of the phylogenetic history of the species. The dogma that depression is a brain disease serves the interests of drug companies, no doubt; it also serves to destigmatize the struggle with depression, which is a good thing. But it is false.

To move forward in our understanding of consciousness, we need to give up the internal, neural microfocus (as Susan Hurley and I once described it). The locus of consciousness is the dynamic life of the whole, environmentally plugged-in person or animal. Indeed, it is only when we take up this holistic perspective on the active life of the person or animal that we can begin to make sense of the brain's contribution to conscious experience.

This is a positive book. Human experience is a dance that unfolds in the world and with others. You are not your brain. We are not locked up in a prison of our own ideas and sensations. The phenomemon of consciousness, like that of life itself, is a world-involving dynamic process. We are already at home in the environment. We are out of our heads.

I have written this book with a particular audience in mind. I imagine that my reader is a lover of science and that he or she is fascinated by the problem of mind, by the fact of consciousness, and by how daunting it is to understand or explain these phenomena. I hope that cognitive scientists and philosophers interested in the mind will read the book and take note of its arguments. But I haven't directed my writing to them. My subject matter is basic to the conduct of normal neuroscience and psychology; it concerns what philosophers call the foundations of cognitive science. I want us to rethink what scientists have simply taken for granted: the basic, starting-point assumptions. For this reason I have tried, in writing this book, to avoid the jargon and insider-speak, the styles of language and argumentation, that already presuppose that one is a member of the cognitive science club.

I am not someone who disdains specialization and technical

language. Science and philosophy are, if you like, conversations that have been going on for a long time. Of course, it will be hard for an outsider to sit down at the table and have a real sense of what is going on. And why should scientists be required to begin again anew each day so that the novice can understand what is under discussion?

The situation is different, however, if the conversation is, well, troubled. In my view, this is the case in contemporary cognitive science. The science of mind could benefit from interruption. It is time to slide our chairs back from the table and to invite intelligent latecomers to join our circle. In cognitive science, specialist jargon and technical details are too often an impediment to clear and honest thinking.

In some sense, then, this book is political. I am writing the book to change the world. Or at least to shake up the cognitive science establishment. I am aware that that's a tall order and that in some ways it may seem presumptuous even to try.

My book is political in another sense as well. American and European intellectual life is fragmented. Humanists—and I don't just mean college English professors but rather anyone whose first love is literature and art—have an awkward relation to science. For many humanists, science is a world apart. Some of them accept its findings uncritically and with indifference. Others disdain science; as far as they are concerned, science has nothing to teach us about what matters most: truth, beauty, art, meaning, experience. Scientists, for their part, have a no less problematic relation to the arts and humanities. Many of them do not appreciate the value of nonscientific research. And those who take an interest in art and literature are very often motivated to explain these phenomena away—for example, by investigating the underlying neuronal basis of aesthetic experience. (This conflict takes a curious and disturbing form where religion is concerned. On the one hand, some religious thinkers hold that religion is altogether insulated from criticism by science, whereas others promote religious doctrine by pretending it is science. On the other hand, scientists, or at least representatives of the scientific worldview, act as though religious people are simply in error, as if they don't realize that religious doctrines lack empirical support.)

In this book I try to show, by example, that science and humanistic styles of thinking must engage each other. Physics used to be called natural philosophy (that's how Newton thought of it). In Germany today, the study of literature is known as *Literaturwissenschaft* (literary science). The idea that science and philosophy, or the humanities more generally, are separate spheres with their own standards and criteria is itself a bit of questionable ideology, a relic of the enthusiasm of an earlier modern age. Natural science is not sui generis. It is not value neutral. It is not discontinuous with broader human concerns. Nor is philosophy a free-for-all of opinion. Philosophy and science share a common aim: understanding. Science and philosophy must work together to advance toward understanding. This is especially so where the target of understanding is consciousness or, more basically, our own nature. The contemporary science of consciousness, at least as it is carried on in the mainstream, rests on shaky philosophical foundations. This makes for an alienated and distorted conception of our human life. It also makes for bad science.

In this book I argue that mind science, like biology more generally, must give pride of place to the whole, living being. I leave it to the reader to judge whether I am successful.

A note about the text of this book. I have made no use of footnotes or in-text references. Instead, I give references or make comments on the text in notes at the end of the book. Each chapter begins with a brief paragraph outlining the aim and topic of the chapter and ends with a brief conclusion or summary.

OUT OF OUR HEADS

AN ASTONISHING HYPOTHESIS

The human body is the best picture of the human soul.
—Ludwig Wittgenstein

Contemporary research on consciousness in neuroscience rests on unquestioned but highly questionable foundations. Human nature is no less mysterious now than it was a hundred years ago. If we are to understand our human nature, we need to make a fresh start. In this first chapter I lay out the basic challenge.

Consciousness Is Like Money

Stop and notice that you can believe in consciousness—appreciate the fact that we feel and think and that the world shows up for us—without believing that there is a place, or a moment in time, when and where consciousness happens or comes to be *inside* of us. As a comparison, consider that there's nothing about this piece of paper in my hand, taken in isolation, that makes it one dollar. It would be ludicrous to search for the physical or molecular correlates of its monetary value. The monetary value, after all, is not intrinsic to the piece of paper itself, but depends on the existence of practices and conventions and institutions. The marks or francs or pesos or lire in your wallet didn't change physically when, from one day to the next, they ceased to be legal tender. The change was as real as it gets, but it wasn't a physical change in the money.

Maybe consciousness is like money. Here's a possibility: my consciousness now—with all its particular quality for me now—depends not only on what is happening in my brain but also on my history and my current position in and interaction with the wider world. It is striking that the majority of scientists working on consciousness don't even notice there is an overlooked theoretical possibility here. They tend to think that consciousness, whatever its ultimate explanation, must be something that happens somewhere and sometime in the human brain, just as digestion must take place in the stomach.

According to the now standard view, our conscious lives—the fact that we think and feel and that a world shows up for us—is achieved in us by the action of our brain. The brain produces images of the environment and manipulates those images in a process known as thought. The brain calculates and infers and eventually produces neural commands so that we act. We really are our brains, and our bodies are at most robotic tools at our brains' disposal. The brain is sole author of what is in fact a grand illusion: that we inhabit a richly detailed and meaningful world, that we are the sorts of beings we think we are. What are we, then? If the truth be told, we are brains in vats on life support. Our skulls are the vats and our bodies the life-support systems that keep us going.

Or so mainstream neuroscience, and writers of science fiction, would have it. Is my body a robot that my brain inhabits? Is the world a grand illusion? Is this really an intelligible conception of ourselves?

Are You Your Brain?

The fundamental assumption of much work on the neuroscience of consciousness is that consciousness is, well, a neuroscientific phenomenon. It happens inside us, in the brain.

All scientific theories rest on assumptions. It is important that these assumptions be true. In this book I will try to convince you that this starting assumption of consciousness research is badly mistaken. Consciousness does not happen in the brain. That's why we have been unable to come up with a good explanation of its neural basis.

Francis Crick, the Nobel Prize–winning codiscoverer of the structure of the DNA molecule, has proposed (in a book titled *The Astonishing Hypothesis*) that "you, your joys and your sorrows, your memories and your ambitions, your sense of personal identity and free will, are in fact no more than the behavior of a vast assembly of nerve cells and their associated molecules." With a flourish, he adds, "This hypothesis is so alien to the ideas of most people alive today that it can truly be called astonishing."

What is striking about Crick's hypothesis is how astonishing it isn't. It isn't surprising to be told that there is a thing inside each of us that thinks and feels and wants and decides. This was the view of the seventeenth-century philosopher René Descartes, who held that each of us is identical to an interior something whose essence is consciousness; each of us, really, is an internal *res cogitans*, or thinking thing. And this is the doctrine promulgated by many religious traditions. Of course, the religions, and Descartes himself, didn't teach that that thing inside us that thinks and feels is a part of our body, a bit of flesh, such as the brain. They supposed that it was something immaterial or spiritual, and so, in that sense, that it was something unnatural. How could mere matter—mere meat—achieve the powers of thought and feeling? Such a possibility boggles the mind. It is precisely on this point, and only on this point, really, that today's neuroscientist breaks with tradition. As Patricia Churchland, a prominent philosopher of neuroscience, has written: "The weight of evidence now implies that it is the *brain*, rather than some nonphysical stuff, that feels, thinks, decides."

But what needs to be kept clearly in focus is that the neuroscientists, in updating the traditional conception of ourselves in this way, have really only succeeded in replacing one mystery with another. At present, we have no better understanding of how "a vast assembly of nerve cells and their associated molecules" might give rise to consciousness than we understand how supernatural soul stuff might do the trick. Which is just to say that the you-are-your-brain idea is not so much a working hypothesis as it is the placeholder for one.

Consciousness researchers in neuroscience like to think that they have broken with philosophy. They have left it behind and set off on the path of science. As Crick has written: "No longer need one spend time attempting . . . to endure the tedium of philosophers perpetually disagreeing with each other. Consciousness is now largely a scientific problem."

Crick is right that the problem of consciousness is now a problem for science. But this doesn't mean that it is no longer a problem for philosophy. For one thing, the aims of philosophy and of science are not different: to achieve understanding of the problems that matter to us. But that's just the beginning: it is a mistake to think that the new neuroscience of consciousness has broken with philosophy or moved beyond it. In fact, as we have been discovering, Crick and other neuroscientists have simply taken a specific family of philosophical assumptions for granted, so much so that their own reliance on them has become all but invisible to themselves. But the fact of the reliance is everywhere in evidence. Its perturbing influence is felt in the seeming mandatoriness of what we can think of as a kind of "gastric juices" conception of consciousness—that is, the idea that consciousness happens in the head the way digestion happens in the stomach. I mentioned before that it is overoptimistic to think of the new neuroscience of consciousness as in its infancy. Developmentally, it would be more apt to characterize it as like a teenager. Like teenagers, neuroscience is in the grip of technology; it has a grandiose sense of its own abilities; and it is entirely lacking a sense of the history of what, for it, seems so new and exciting.

A Really Astonishing Hypothesis

It *would* be astonishing to learn that you are *not* your brain. All the more so to be told that the brain is not the thing inside of you that makes you conscious because, in fact, there is no thing inside of you that makes you conscious. It would then turn out that contemporary neuroscience has been in the thrall of a false dichotomy, as if the only alternative to the idea that the thing inside you that thinks and feels is immaterial and supernatural is the idea that the thing inside you that thinks and feels is a bit of your body. It would be astonishing to be told that we've been thinking about consciousness the wrong way—as something that happens in us, like digestion—when we should be thinking about it as something we do, as a kind of living activity.

In this book I advance this truly astonishing hypothesis: to understand consciousness in humans and animals, we must look not inward, into the recesses of our insides; rather, we need to look to the ways in which each of us, as a whole animal, carries on the processes of living in and with and in response to the world around us. The subject of experience is not a bit of your body. You are not your brain. The brain, rather, is part of what you are.

A Note on Terminology, and the Thesis Restated

In this book I use the term "consciousness" to mean, roughly, experience. And I think of experience, broadly, as encompassing thinking, feeling, and the fact that a world "shows up" for us in perception. Many writers have sought to define terms more narrowly than this. No doubt there are important distinctions that can and, for certain purposes at least, should be drawn. For example, a contrast is often made between thought or cognition, on the one hand, and sensation and feeling, or phenomenal experience, on the other. The contrast is between planning and carrying out an action, for example, and, say, experiencing the taste of licorice. When people draw this distinction it is usually because they think it is much easier to explain thought, say, than it is to explain the quality of our conscious episodes. For example, many theorists hold that thinking is a matter of computation and that we shed light on how brains think by comparing them with computers. As I discuss in Chapter 7, it is far from true that computers can think; moreover, I argue there, computers can't think largely for the same reason that brains can't. Meaningful thought arises only for the whole animal dynamically engaged with its environment, or so I contend. And indeed the same is true for the quality of our conscious episodes. The taste of licorice is not something that happens in our brains (although it is true that when we eat licorice, we do so by putting it in our mouths).

Conscious states are typically states that I can talk about, that influence what I do, and so they are states that I can make use of in planning. For example, my dislike of the taste of licorice is something that informs my larger cognitive and behavioral life. Among other things, it will influence my shopping behavior. Such a state is available or accessible to thought and talk; it is sometimes said that this marks a distinctive sort of consciousness that the philosopher Ned Block has named access conscious-ness. The access consciousness of my feelings about licorice is one thing, however, and the experience of the licorice itself is another. The latter is an episode in what Block has called phe-nomenal consciousness, and the question of whether an episode is phenomenally conscious is, or so it seems, altogether different from the question of whether it is access conscious. To ask whether an episode is phenomenally conscious is to ask, in the philosopher Thomas Nagel's phrase, whether "there is some-thing it is like to be" in that state. To ask whether it is access con-scious is to ask whether the occurrence of the state influences what we say and do and want and plan and so on.

Other distinctions abound. To be conscious, as opposed to be-ing unconscious, is to be awake, aroused, alert, as opposed to being asleep or knocked out. In ordinary language, self-consciousness means a kind of interfering attentiveness to how others view oneself. In philosophy and cognitive science, self-consciousness means something different. Self-consciousness is that feature of experience by virtue of which our experiences are *ours*. Experi-ences have a kind of "mine"-ness that makes them, distinctively, our own, or so some thinkers have maintained. Freud famously hypothesized the importance of unconscious desires and wishes in explaining human psychology.

Distinctions are useful, depending on your purposes. For my purposes, these distinctions don't matter in particular. When they do, I'll try to be careful to be clear about what I am refer-ring to. The problem of consciousness, as I am thinking of it here, is that of understanding our nature as beings who think, who feel, and for whom a world shows up.

Another terminological issue arises in connection with the words "mind" and "brain." The latter refers to a part of the body found in the head and connected up to a larger system known as the nervous system. It is widely believed that the brain and the larger nervous system of which it is a part play a special role in explaining our powers of mind (e.g., thought, memory, perception, emotion, and the like). Indeed, some scientists and philosophers think that the mind *is* the brain. Be that as it may, it is important to realize that no one holds that the concept of brain and the concept of mind are the same. To have a mind is, roughly, in my sense, to be conscious—that is, to have experience and to be capable of thought, feeling, planning, etc. To have a brain, on the other hand, is to have a certain kind of bodily organ or part. Ordinary language is sometimes a bit confused about this, so we need to be careful. Being intelligent, for example, is said to be a matter of having brains.

My central claim in this book is that to understand consciousness—the fact that we think and feel and that a world shows up for us—we need to look at a larger system of which the brain is only one element. Consciousness is not something the brain achieves on its own. Consciousness requires the joint operation of brain, body, and world. Indeed, consciousness is an achievement of the whole animal in its environmental context. I deny, in short, that you are your brain. But I don't deny that you have a brain. And I certainly don't deny that you have a mind. To have a mind, though, requires more than a brain. Brains don't have minds; people (and other animals) do.

The Man with Two Brains

I have always been a fan of Carl Reiner's hilarious movie *The Man with Two Brains*. Steve Martin plays the lead role, a brain surgeon named Dr. Hfuhruhurr who falls in love with a disembodied brain awaiting a transplant. This is the brain of the woman of his dreams. Now all he needs—all *she* needs—is a body! He sets about a villainous scheme to get his hands on the body of the beautiful and svelte Dolores Benedict (played by Kathleen Turner). The joke is that, unbeknownst to him, the self whose brain he loves has an eating disorder. By the time she recovers from the brain transplant, she's morbidly and unattractively obese. (He loves her anyway!)

This is the stuff of science fiction. Pretty far-fetched, to be sure. The fact that we find it at all comprehensible, let alone compelling, shows that the "astonishing hypothesis" of the establishment neuroscientist now belongs to the conventional wisdom of the culture at large. We think of ourselves—or find it easy to take seriously the idea of ourselves—as dependent on our brains in a special sort of way, very different from the way we depend on our hearts, say. You gotta have heart, yes. But it is the brain, with its distinctive neuronal snap, crackle, and pop, that is our ground. We inhere in our brains. What makes us the kind of thing we are—beings who can feel and reason and think and see—is accomplished in our bodies by our brains.

I ask again: Is this a plausible conception of ourselves? Reiner's movie casts an interesting light on this question. The film itself needs to present us with communication between the Martin character and his beloved brain-in-a-cookie-jar. But how can it do this? How, for example, to capture the fact that the lovely female voice Martin hears—what we in the audience experience as a voice-over—is actually the voice of the person in that brain-in-a-cookie-jar? Film normally trades on the ventriloquist effect. We hear the voice coming from the mouth because we see the mouth move in synchrony with just those words. Vision captures and directs what we hear. In fact, this is an important part of normal speech perception. The problem with a speaking brain is that it has no mouth. What ties the sounds to the brain? What makes them *its* words? The movie strikes on a silly but funny solution. The brain glows and pulsates in synchrony with its spoken words. What makes this solution interesting, as well as silly and funny, is that, in a way, it's cheating. Brains don't pulsate or change colors, and by introducing this feature you are, in effect, giving the brain a body or, more important, a face (what the brain is supposed to lack). And maybe that's not just a somewhat confused filmic conceit but something of a conceptual necessity. It's hard even to conceive of a consciousness that lacks a face. That's why, tragically, even friends and family find it difficult to empathize with Parkinson's patients whose faces have grown masklike. And that's why, in a love scene in *The Man with Two Brains*, the Steve Martin character puts a scarf around the base of his love's brain-in-a-cookie-jar, a hat on top, and bright red candy-wax lips on the front. Wittgenstein wrote that it is only of what looks and behaves like a person that we say it sees, thinks, feels. The problem with a brain is that it doesn't look and behave like a person.

Consciousness in a Petri Dish?

If the new neuroscience establishment is right, then it ought to be possible, at least in principle, to have consciousness in a petri dish. All that would be required for consciousness in a petri dish is that the cells be wired up to each other and stimulated in a suitable matter.

My own view is that the suggestion that cells in a dish could be conscious—or that you could have a conscious brain in a vat—is absurd; it's time to overhaul our starting assumptions about what consciousness is if they lead us to such a conclusion.

Consider, first of all, that the vat, or petri dish, couldn't be a mere dish or bucket, as Evan Thompson and Diego Cosmelli have discussed in an essay. It would have to supply energy to nourish the cells' metabolic activity and it would have to be capable of flushing away waste products. The vat would have to be very complicated and specialized in order to control the administration of stimulation to the brain comparable to that normally provided to a brain by its environmentally situated body. If you actually try to think through the details of this thought experiment—this is something scientists and philosophers struck by the brain-in-a-vat idea almost never do—it's clear that the vat would have to be, in effect, something like a living body. But then, it would seem, the thought experiment teaches us what we knew already: not that we are our brains but rather that living animals like us can be, well, conscious.

Presumably it is an empirical question just how many cells would be necessary for conscious activity. It is consistent with what we now know to think it could very well turn out that in order to get consciousness in a vat, you'd have to have a whole, suitably activated, healthy brain in the vat. Recent work on the neural basis of visual consciousness has, as a matter of fact, tended to suggest that large-scale, ongoing interactions between widely separate areas of the brain are necessary for visual consciousness.

But now ask yourself: Do we have any reason, in advance of our Frankensteinian researches, to think that the whole brain is the outer limit of what might be needed for consciousness in a petri dish? If we can't draw the boundary in advance at this or that brain region, then how can we be confident that we can draw the boundary at the limits of the brain itself? Maybe consciousness depends on reliable interactions between what is going on in the brain and what is going on in nonbrain parts of the body. It could even turn out that consciousness depends on interactions between the brain and the body and bits of the world nearby. So maybe, to get consciousness in the dish, we'd need not only brain and body but also a reasonable facsimile of the environment in the dish too.

The point of this line of questioning will by now be clear. Our philosopher-neuroscientists with their brain-in-a-vat fantasies fail to notice that their fantasies are taking a stand on what surely is an open empirical question: How much is minimally necessary for consciousness in a petri dish?

Taking the Problem Seriously

These are hard questions. And they aren't merely academic. Consider the case of a thirty-nine-year-old Belgian stroke victim who fell into a coma. Laura Spinney in the *Guardian* (April 15, 2004) reported:

> Doctors concluded that she was unlikely to regain consciousness and, after a time, diagnosed her condition as persistent vegetative state (PVS). One of the criteria on which they based their decision was her inability to blink or track a moving object with her eyes. It was only when they discovered that the stroke had damaged a cranial nerve, preventing her from opening her eyes, that they realized their error. If they opened her eyes for her, she followed their instructions. Having regained full consciousness soon after her stroke, she revealed she had overheard all the bedside discussions as to whether it was worth keeping her alive. At no point had she wanted to die.

The misdiagnosis of persistent vegetative state is horrifying but all too understandable. In normal circumstances it isn't difficult to know whether someone is uncomfortable, or in pain. In normal circumstances, how we feel finds expression in our faces and in our movements. These movements of the face, voice, and body are not mere signals to others, devices for effectively communicating with them. We don't first feel glad and then choose to express our gladness to others in a smile, just as we don't first feel pain and then produce a grimace for the information of others. As William James first noticed, the grimace and the smile belong to our state of consciousness. They are not so much evidence of what is going on within us as they are, in fact, enactments of our condition. They are its natural expression. And there are probably good evolutionary reasons for this. The fright my fellow human being (or monkey or chimpanzee or whatever) feels at the arrival of a predator is not of much less significance to me than it is to him, and group cohesion surely depends on our ability to read each other's mind.

The point is that circumstances are not normal in the clinic. Obviously, the mere absence of the normal behavioral markers of consciousness does not entail the absence of consciousness. But what is the alternative to looking and listening to what someone says and does, to how they look? The Belgian stroke patient was lucky twice over. First, her mental presence was in fact detected. Second, she quickly recovered. Other patients with severe paralysis and loss of speech have not been so lucky. For example, thirty-two-year-old Julia Tavalaro spent six years in a New York chronic care hospital where she was known as "the vegetable" before a loved one noticed indicators of consciousness. In fact, she was entirely conscious the whole time: she was simply unable to give any signs to others. She'd spent six years trapped inside an inert body, unable to communicate with the outside world in any way. She eventually returned home and died at the age of sixty-eight. This condition, now known as locked-in syndrome, is brought about by brain-stem injury typically caused by a stroke. Because of the anatomy of the brain stem, patients with "classical" locked-in syndrome are typically able to move their eyes and deploy elaborate blinking and looking codes to communicate. Several such patients have written books. I have seen a videotape of a man with locked-in syndrome. In the initial shots you see the impassive and inert face of a man who appears to be blinking reflexively. The camera slowly pans back and you realize that the man is in fact staring at a computer screen. With his blinking he is actively controlling a cursor on the screen and managing an online database of locked-in syndrome sufferers in France!

But there are also known cases of total locked-in syndrome. The correct diagnosis of total locked-in syndrome—or, indeed, the more typical eye-movement locked-in syndrome—is extremely difficult. Tellingly, family members or caregivers are more likely to make the diagnosis than physicians. Sadly, it is almost certain that until recently all patients with locked-in syndrome have been mistakenly supposed to be mere vegetables, lacking all sentience, and have probably been allowed to endure slow and painful deaths by starvation. There are very few documented cases of total locked-in syndrome. This in itself is a frightening fact.

One does not need to turn to extreme forms of brain injury such as locked-in syndrome to appreciate the practical importance of the problem we are isolating. When my four-year-old son August was in the hospital for a hernia operation, before they wheeled him into surgery I asked the anesthesiologist whether he could assure me that August would not suffer any pain or discomfort during the operation. He replied that there was no cause for worry: he would personally monitor August's heart rate and would watch his face closely for signs of discomfort. I was reassured that the doctor would be paying attention. But I certainly wondered whether the absence of these very primitive behavioral and physiological indicators was reliable evidence that my son was free of awareness of what he was undergoing.

Locked-in syndrome, and the medical practice of anesthesiology, are forceful reminders that doctors can't afford to rely

alone on behavioral expressions of mental state. Persistent vegetative state, in contrast, serves to remind us of the converse. The
persistent vegetative state is thought to be a condition of wakefulness without consciousness. But it is not uncommon for patients in this condition to respond to sounds, to sit up and move
their eyes, to shout out, to grimace, to laugh, smile, or cry. Suppose it is your beloved who lies there jumping at the sound of
the door slamming, her eyes darting around furiously. She cries
out in seeming rage or purrs with apparent contentment. What
would convince you that your loved one is unfeeling, absent, that
she has become a vegetable? Whereas with locked-in syndrome
we are challenged to believe that behind the masklike wall of a
face there is a lively intellect at work, with persistent vegetative
state we struggle to take seriously the thought that there is an absence of feeling and subjectivity behind what moves us as an ex-
pressive face.

Looking into the Head

We might turn to the technologies of brain scanning in the hope
that these will enable us to look into the living brain itself to find
out what is going on in there. The fact that brain-imaging studies of patients with locked-in syndrome—positron emission
tomography (PET), functional magnetic resonance imaging
(fMRI), as well as electroencephalography (EEG)—tend to show
normal levels of cortical activity can be taken to be a confirmation of the judgment that patients with locked-in syndrome have
normal mental lives. It is much harder, though, when we turn to
patients in the persistent vegetative state. Here what confronts
us is not so much direct evidence of the lack of consciousness
as the absence of normal brain-imaging findings. Does the absence of normal brain profiles in patients in the persistent vege-

tative state help us decide whether they are sentient or not? Would the mere absence of normal patterns of neural activity as modeled by functional imaging technologies such as fMRI or PET satisfy you that your loved one was now little more than a vegetable?

Actually, things are more complicated. Although patients in the persistent vegetative state show markedly reduced global brain metabolism, so do people in slow-wave sleep and patients under general anesthesia. But sleepers and surgery patients wake up and resume normal consciousness, whereas patients in the persistent vegetative state rarely do. Remarkably, in the small number of cases in which brain imaging has been attempted in patients who have recovered from the persistent vegetative state, regaining full consciousness, it would appear that global metabolic levels remain low even after full recovery. Moreover, external stimuli such as sounds or pinpricks produced significant increases in neuronal activity in primary perceptual cortices. Interesting new work by Steven Laureys and his colleagues in Belgium indicates that vegetative patients show strikingly impaired functional connections between distant cortical areas and between cortical and subcortical structures. In addition, they show that in cases where consciousness is recovered, even if overall metabolic activity stays low, these functional connections between brain regions are restored. These findings are important and point in the direction of a deeper understanding of what is happening in the brain in the persistent vegetative state.

But this doesn't change the fact that at present we are not even close to being able to use brain imaging to get a look inside the head to find out whether there is consciousness or not. Consider these simple questions: Does a patient in the persistent vegetative state feel physical pain—for example, the pain of thirst or hunger, or the prick of a pin? Does she hear the sound of the door slamming? We know she turns her head in response to the sound, and we know she withdraws her hand from the pinprick. We also know that there is some significant neural activity produced in primary perceptual cortices by these stimuli. Is the patient in the persistent vegetative state a robot, responding reflexively to stimulation, but without actually feeling anything? And, more important, is this something that brain imaging could ever help us decide?

We don't know how to answer these questions. It is disturbing to learn that so far there are no theoretically satisfying or practically reliable criteria for deciding when a person with brain injury is conscious or not. At present, doctors and relatives have to deal with these questions without guidance from science or medicine. For example, the press tended to treat the widely discussed case of Terri Schiavo as one in which science, armed with cold hard facts about the nature of Schiavo's brain damage, did battle with family members who were blinded at once by their love for their daughter and their religious fundamentalism. Sadly, science doesn't have the hard facts.

The New Phrenology?

It would be hard to overstate the extent to which the fervor about the brain-based view of consciousness is driven by the development in the last few years of new technologies of brain imaging. Until very recently, postmortem autopsy has been just about the only way to examine the brain of a person with known neurological deficits. Ethical considerations prevent scientists from deploying the sorts of invasive techniques that are used on animals. The brain has remained, for science, a black box. At best we have been able to draw conclusions about its design and functionality by looking at what possessors of brains can say and do. Things are different now, or so it is widely believed. The development of PET and more recently fMRI—technologies of functional imaging—now enable us to penetrate the black box. Brain imaging provides colorful pictures of the brain, enabling us to see how it lights up in action as it performs its functions.

Given the huge personal and institutional investment in brain-scanning methods and technologies, it is understandable that there is so much hype about the power of functional imaging. It is hard to doubt that these technologies will add to our ability to move forward in our quest to understand the conscious mind. But this is all the more reason to pause and step back from the hype. In fact, functional imaging raises important and still unresolved methodological problems.

PET and fMRI yield multicolored images. The colors are meant to correspond to levels of neural activity: the pattern of the colors indicates the brain areas where activity is believed to occur; brighter colors indicate higher levels of activity. It is easy to overlook the fact that images of this sort made by fMRI and PET are not actually pictures of the brain in action. The scanner and the scientist perform a task that is less like gathering a photographic or X-ray image than it is like the process whereby a police sketch artist produces a drawing of a suspect based on interviews with a number of different witnesses. Such drawings carry valuable information about the criminal, to be sure, but they are not direct records of the criminal's face; they are, rather, graphical renderings based on perhaps conflicting reports of what different individuals claim to have seen. Such a composite sketch reflects a conjecture or hypothesis about, rather than a recording of, the perpetrator. Indeed, there is nothing in the process that even guarantees that there is a single perpetrator, let alone that the sketch is a good likeness.

In a similar way, images produced by PET and fMRI are not in any straightforward way traces of the psychological or mental phenomena. Rather, they represent a conjecture or hypothesis about what we think is going on in the brains of subjects. To appreciate this, consider that we face a problem from the very beginning about how to decide what neural activity is relevant to a mental phenomenon we want to understand. Scientists start from the assumption that to every mental task—say, the judgment that two given words rhyme—there corresponds a neural process. But how do we decide which neural activity going on inside you when you make a rhyming judgment is the neural activity associated with the mental act? To do that, we need to have an idea about how things would have been in the brain if you hadn't performed the rhyming judgment; that is, we need a baseline against which to judge whether or not the deviation from the baseline corresponds to the mental act. One way to do this is by comparing the image of the brain at rest with the image of the brain making a rhyming judgment. The rhyming judgment presumably depends on the neural activity by virtue of which these two images differ. But how do we decide what the brain at rest looks like? After all, the brain is never at rest. For example, there are stages of sleep when your brain is working harder than it does at most times during the day!

Comparison provides the best method available for uncovering the areas of the brain that are critically involved in the performance of a cognitive function. For example, suppose you were to produce a bunch of PET images of people listening to recordings of spoken words and then making judgments about whether given pairs of words rhyme. To isolate the activation responsible for the rhyming judgment, as distinct from that responsible for the auditory perception of the spoken words, a standard procedure would be to compare these images with a second set of images of people listening to recordings of spoken words but not making rhyming judgments. Whatever areas are active in the first set of images, but not the second, would be plausible candidates for the place in the brain where the rhyming judgment takes place.

This method of comparison is cogent and it holds promise. But it is worth stressing that its reliability depends on a number of background assumptions, not all of which are unproblematic, as Guy C. Van Orden and Kenneth R. Paap have convincingly argued. For one thing, sticking to our example, the comparison method assumes that there is no feedback between what the brain is doing when we make a rhyming judgment and what the brain is doing when we perceive the words. If there is indeed feedback, then it would follow that overlapping regions in the images do not necessarily correspond to a common neural factor.

As a matter of fact, it is highly likely that there is feedback. Neural activity in the brain during perception, for example, is not a one-way thing. Neural activity is characterized by loops and two-directionality. There are neural pathways heading back into the brain from the senses, but there are even more neural pathways heading back out again. This should not be surprising. Consider how much easier it is to hear a sound that you are expecting than one that you are not expecting. The assumption that there is no feedback in the neural circuitry is the flip side of a different assumption that we can factor the cognitive act itself into distinct, modular acts of perceiving the words (on the one hand) and judgments about whether they rhyme (on the other). That's a substantive empirical claim about the character and composition of cognitive acts themselves and certainly not something that can be simply taken for granted.

I am using the rhyming case as an illustrative example. My aim is not to show that, in fact, the method of comparison is misguided. What I do want to bring out is that brain scanners don't simply show us what is going on when we listen and judge. In a way, these considerations about feedback in the brain and cognitive models are only the tip of the iceberg. PET and fMRI have very low spatial and temporal resolution. When you localize events in the brain using these techniques, you localize them to cubic regions of between 2 and 5 mm—that is, to regions in which there are hundreds of thousands of cells. If there is specialization or differentiation among these cells, that won't show up in the picture. Nor, for that matter, can we be sure exactly when neural events are happening. Cellular events unfold at the scale of thousandths of a second, but it can require much longer time scales (large portions of a minute) to detect and process signals for making images. For these reasons, scientists have developed techniques of normalizing data. Typically, data from different subjects is averaged. The averaging process involves the loss of considerable information. After all, brains differ from one another no less than faces and fingerprints do. Just as the average American taxpayer has no set height and weight, so averaged neural activity has no set location in any particular brain. For this reason, scientists project their findings onto an idealized, stock brain. The pictures we see in the science magazines are not snapshots of a particular person's brain in action.

Finally, putting all this to one side, it is important to be clear that there is no sense in which PET or fMRI pictures deliver direct information about consciousness or cognition. They do not even deliver direct representations of neural activity. Functional brain-imaging systems such as PET and fMRI build images based on the detection of physical magnitudes (such as radio or light waves) that are believed to be reliably correlated with metabolic activity. For example, in PET, one injects a positron-emitting isotope into the bloodstream; PET detects the emission of gamma rays caused by the collision of positrons and electrons. In this way, the PET image carries indirect information about metabolic activity based on the direct measurement of a physical magnitude, which is in turn supposed to carry information about neural activity. The latter supposition is not unreasonable. Neural events require oxygen, and so they require blood. The neural activity, in its turn, is supposed to correlate to significant mental activity. Brain scans thus represent the mind at three steps of removal: they represent physical magnitudes correlated to blood flow; the blood flow in turn is correlated to neural activity; the neural activity in turn is supposed to correlate to mental activity. If all the assumptions are accurate, a brain-scan image may contain important information about neural activity related to a cognitive process. But we need to take care not to be misled by the visual, pictorial character of these images. Brain scans are not pictures of cognitive processes in the brain in action.

Conclusion: You Are Not Your Brain

Empirical research on consciousness and human nature takes for granted that the problem for science is to understand how consciousness arises in the brain. That consciousness arises in the brain goes unquestioned. In the meantime, guns blazing, engines roaring, we are going nowhere in our quest to understand what we are. In this chapter I ask whether our inability to explain consciousness and the workings of our minds stems precisely from our unquestioned assumptions. In the remainder of this book I seek to demonstrate that the brain is not the locus of consciousness inside us because consciousness has no locus inside us. Consciousness isn't something that happens inside us: it is something that we do, actively, in our dynamic interaction with the world around us. The brain—that particular bodily organ—is certainly critical to understanding how we work. I would not wish to deny that. But if we want to understand how the brain contributes to consciousness, we need to look at the brain's job in relation to the larger nonbrain body and the environment in which we find ourselves. I urge that it is a body- and world-involving conception of ourselves that the best new science as well as philosophy should lead us to endorse.

2

CONSCIOUS LIFE

My attitude towards him is an attitude towards a soul. I am
not of the *opinion* that he has a soul.

—Ludwig Wittgenstein

I begin with what can seem to be the most challenging of problems about consciousness, what philosophers call "the problem
of other minds." Can we know the minds of others? How do we
decide whether other people are conscious? And what about the
consciousness of other species? The problem of other minds can
seem insurmountable. This is because we think that the problem
we face is a theoretical one: how to acquire knowledge of another's mind on the basis of what he or she says and does, or on
the basis of a neural signature. But we don't face this problem.
The basis of our confidence in the minds of others is practical.
We cannot take seriously the possibility that others lack minds
because doing so requires that we take up a theoretical, detached stance on others that is incompatible with the kind of
life that we already share with them. All this points to something
paradoxical about the science of the mind: science requires detachment, but mind can only come into focus if we take up an
altogether different, more engaged attitude. Does this mean a
science of the mind must be impossible? No. There is a way forward for science. The solution comes when we recognize that
there is a rigorously empirical alternative to mechanistic detach-

ment on the one hand and mere personal intimacy on the other. This is the perspective of biology.

Other Minds

Who, or what, is conscious? How can we decide? Where in nature do we find consciousness? This problem can seem like the hardest one in this whole field: the question of the consciousness of others. I am aware. So are you. We think, we feel, the world shows up for us. But what about an ant, or a snail, or a paramecium? What about a well-engineered robot? Could it be conscious? Is there a way of telling for sure?

The start of almost all reflection on this problem is the idea that our knowledge of how others think and feel—indeed, our knowledge that they think and feel and are not mere automata— is based on what we can see and hear and measure. We observe behavior or, as in the case of patients in the persistent vegetative state or with locked-in syndrome, we measure neural activity. It can seem, then, that the closest we can come to knowing other minds, in a theoretically respectable way, is having some account according to which behavior and neural activity provide reliable criteria of a person's psychological state.

But this admission is really to concede that we don't have knowledge of other minds, at least not in a scientifically respectable way. For observations of behavior (what people say and do) and measurements of neural activity don't yield knowledge of other minds. Surely this is an important lesson from the persistent vegetative state and locked-in syndrome. Experience need not be reflected in what we say and do; mere behavior is at best an unreliable guide to how things are for a person. Moreover, we really don't understand the connection between neural activity and experience anyway. As I stressed in Chapter 1, it is entirely unclear whether results of a brain scan would ever or ought ever to convince us that our daughter was no longer a living person, especially when she continues at least to appear to respond to sounds and touch. If what people say and do, and measurements of what their brains are doing, are the best we have to go on, then it would seem that our commitment to the minds of others is philosophically ungrounded, a mere act of faith.

Perhaps we have simply evolved to see mind and consciousness in the world around us, even when there is none. Humans do seem to have a rather striking propensity to personify and animate what is mechanical and lifeless. The psychologists Fritz Heider and Marianne Simmel back in the 1940s produced a startling example of this. They made an animation in which simple geometrical figures—a small circle and a larger and a smaller triangle—moved about on a screen.

Heider and Simmel found that ordinary subjects, when asked to describe what was on display in the film, tended to "anthropomorphize" the shapes—that is, to attribute to them genders and also, more important, goals, intentions, and mental attitudes such as fear. Subjects viewed the movements of the shapes as forming a narrative in which, for example, a threatening triangle was chasing a circle while a small triangle tried to intervene on the circle's behalf. Heider and Simmel demonstrated that most people viewing this meaningless cartoon interpret what they are seeing "in terms of actions of animated beings, chiefly of persons."

Or consider Kismet, a robot designed by Cynthia Breazeal and her colleagues in Rodney Brooks's lab at MIT. Breazeal's aim is to design robots that are easy to use and that can interact with people in normal "social exchange." To this end, the robot is designed to detect objects of interest (faces or toys) and to "express" a range of attitudes and emotions (e.g., happiness, sadness, surprise, pleasure, boredom, anger, calm, displeasure, fear, and interest), depending on what it "perceives." Kismet is also equipped with a voice synthesizer: although she is not built to understand anything, she will respond to conversation with conversationally appropriate sounds and tones, though without words.

Now, we all know that it is easy to project feelings onto dolls and pets and dumb objects—some of us are old enough to remember pet rocks!—but what is remarkable is the degree to which interaction with Kismet affords the pleasures of human interaction. The illusion of attention, interest, and presence is so strong that Kismet captures and holds the attention of the graduate students who participate eagerly in extended conversation with her. Of course, Kismet says nothing, feels nothing, knows nothing. She isn't made to fool us, but she's made in such a way that we can't even worry too much about being fooled. It is hard to imagine a more powerful demonstration that we human beings have something like a will to experience mind whether it is there or not.

It is not difficult to think up an evolutionary rationale for the possession of this trait of overly liberal attribution of mind. Better that we have false positives and overattribute mind to puppets and shapes than that we be caught unaware in our ancestral forest. But the question of interest for us is: Is my confidence that *you* are a real locus of thought and feeling any better founded than my belief that Kismet is? Given the impoverished grounds on which we are able to make attributions to others— roughly, physiological indicators of mental activity plus what they say and do—it can come to seem inescapable that what requires explanation is not our knowledge of other minds but rather our entirely ungrounded idea that we can know what another thinks or feels or wants, or even whether another thinks or feels or wants at all.

"Theory of Mind"

Suppose I take the chocolate from the drawer where you left it and put it someplace else. Where will you look for the chocolate when you get home? Children younger than four or five, it turns out, will rarely get this right. When questioned, they will say, mistakenly, that you will look for the chocolate at its new location. They don't appreciate that your actions will be governed by your false belief that the chocolate is where you left it, rather than, as if this were possible, by the chocolate itself. These children, it is said, fail the false-belief test, and this failure in turn is taken to be evidence that they don't yet have a grip on other minds. They perceive behavior, but they have no conception of others as distinct subjectivities who are limited by their own points of view. At about the age of five, it is claimed, children begin to see others as conscious beings; this transition happens when children acquire what is called a "theory of mind"—that is, a conception of the mind as an unobservable domain of forces whose effects are evident in what people say and do. One metaphor sometimes used by researchers working within this framework is that young children are like scientists trying to figure out how to make sense of what they perceive going on around them. The crucial developmental milestone is when they come to appreciate that they can predict and explain the actions of others; actions are vectors with belief and desire as components. According to some scientists, theory of mind is a uniquely human cognitive technology: nonhuman primates—even chimpanzees—never get it; they'll make predications about the actions of another chimpanzee, but they will not do so on the basis of an assessment of what that other chimpanzee sees and believes and wants. Chimpanzees, like young children, are behaviorists, confined to making sense of others only based on what they can observe. And not all human children acquire theory of mind. An influential hypothesis about the sources of autism in children is precisely the thought that these children fail to understand others as, like themselves, capable of feeling, thinking, wanting, and acting out of psychologically potent motivations.

The idea that our conception of other minds is a strategy or stance for predicting and explaining the doings and sayings that we witness can be thought of as a kind of skeptical solution to the problem of other minds. It takes for granted from the start that all that is available to us is the mere behavior of others; it takes for granted that minds are hidden and private. It also takes for granted that the minds of others are real for us only as a kind of theoretical device to help us manage our dealings with others. Just as we posit the existence of an unseen planet to account for perturbations in the orbit of a planet that we do perceive, so we explain why your body will travel along the space-time path that it does by appealing to a domain of unperceived, merely hypothetical causes. In other words, you open the drawer because you want chocolate and have the false belief that that is where the chocolate is.

Personal Intimacy

Now, there is, in fact, ample evidence that young children are highly sensitive to the feelings, attitudes, interests, and purposes of others long before they pass the false-belief test. This finding would be obvious to anyone who has ever spent time in the company of young children, or so I would have thought. Young children participate, actively and purposefully, in very delicate communicative relations with their mothers and other caretakers. Infants respond to the look, smile, voice, and touch of their caretakers; and they are made uncomfortable when that care is withdrawn. To verify this, psychologists have developed the so-called still-face paradigm. They observe the effects on the child of the mother's face becoming still or impassive. What experimental observation demonstrates is that children are distressed by this sort of visible withdrawal on the mother's part. The child will make efforts to recapture the mother's attention and elicit a response from her; if no response is forthcoming, the child will turn away, visibly subdued. This is powerful evidence that children are interested in their mothers and actively participate in emotional exchange.

There is room for controversy here, I suppose. There are those who believe that children merely act as though they appreciate the thoughts and feelings of others—that they are natural robots not altogether unlike Kismet. According to this line of thought, this trait is fortunate for the species. Children act as though they are sensitive to their caretakers, and so, in turn, the caretakers are motivated to take care of their children and enable their cognitive development to take place according to its own internal time clock. But this is not persuasive. Far more reasonable is a different picture according to which the social dynamic of mother and child basically makes up the process whereby the mental life of the child takes form. Children are not separate; they are not observers; they are regulated by their mothers' soothing or alerting tones, eye contact, gestures, and touch. A mother is literally one of the structures constituting a child's psychological landscape.

From this other perspective, the child is always already in a context in which shared feeling and mutual responsiveness are a given. The baby's relation to the other unfolds in an emotional setting. The baby never faces the predicament of needing to figure out that his mother is animate; it is never the case that the child is more intimately acquainted with mere gesture or behavior and that he needs to posit feeling or mind behind that outward screen. And all of this is compatible with the fact that the child of three cannot pass the false-belief test. It may very well be right that young children take for granted that they and others share a common world and also common interest in that world, and that this assumption leads them to be unable to fathom that the attention-grabbing pull of the chocolate itself won't be experienced by others as it is experienced by them. But far from showing that they have no conception of the minds of others, this shows that they have no conception of the minds of others as private and unobservable. And in a way they are right about this.

Accepting Ungroundedness

There is one more piece to the puzzle about our knowledge of other minds. It is this: No sane person can take seriously the suggestion that our knowledge of other minds is merely hypothetical. However weak our evidence that others have minds may be, it is plainly outrageous to suggest that we might, for this reason, give up our commitment to the minds of others. That my wife and children and parents are thinking, feeling beings, that a world shows up for them—that they are not mere automata—is something that only insanity could ever allow me to question. This fact of our moral certainty that those we share life with are themselves conscious beings needs to be explained.

One might simply hold this fact up as further evidence that our commitment to other minds is without justification. It would be almost right to say so. It does show that our commitment to the minds of others is not based on evidence. So in a way it is true that our commitment to the living consciousness of others is ungrounded. But it is not, for this reason, unjustified or misguided. This is because our commitment to other minds is, I would like to propose, not really a theoretical commitment at all. We don't come to learn that others think and feel as we do, in the way that we come to learn, say, that you can't trust advertising. Our commitment to the consciousness of others is, rather, a presupposition of the kind of life we lead together.

In this respect, the young child, in her relation to the caretaker, is really the paradigm. As I have suggested, the child has no theoretical distance from her closest caretaker. The child does not wonder whether Mommy is animate. Mommy's living consciousness is simply present, for the child, like her warmth or the air; it is, in part, what animates their relationship. Mommy's mind and Baby's mind come to be in the coochy-cooing directedness that each sustains toward the other. If one wants to speak of a commitment to the alive consciousness of others here, one should speak not of a cognitive commitment but, rather, of a practical commitment. Like the baby in relation to her mother, we are involved with each other. It is our joint cohabitation that secures our living consciousness for each other. We live and work together.

This line of reasoning, more than anything else, explains why no sane person can take doubts about other minds seriously. From the standpoint of these collaborative mutual involvements, the problem of other minds can't arise: not because there is not in fact a detached, theoretical standpoint from which the problem can be raised—of course there is—but because, crucially, we do not and cannot occupy that standpoint, at least as long as we want to carry on together in cooperation. In a way, this last point is obvious. Louis Armstrong famously said, when asked what jazz is, "If you gotta ask, you ain't never gonna know." Intimacy and commitment sometimes simply leave no room for theoretical musings. I cannot both trust and love you and also wonder whether, in fact, you are alive with thought and feeling, just as I cannot dance well if I am counting steps and trying to remember what comes next. A certain theoretical detachment is incompatible with our joint mutual commitment.

Again, the point of this discussion is not that our commitment to each other's consciousness is beyond rational criticism. We have already considered some painful ways in which the mind of another can be thrown into doubt for us, namely, the persistent vegetative state and locked-in syndrome. The point, crucially, is that for a person's mind to be thrown into doubt for us does not mean we have lost the evidence we once possessed that assured us from a standpoint of theoretical detachment that the other was mentally present. If I am right, that is a standpoint that we never occupy in relation to other minds (or that we occupy only rarely, in special circumstances). Precisely what is thrown into question in cases such as these is the question of what our relation to the other *should* be. This is a practical, or rather, a moral question. That is why the persistent vegetative state is a problem for morality as much as it is a problem for science. When the family of a person who has "become a vegetable" refuses to give up on that person, refuses to consider "pulling the plug" or "pulling out the feeding tubes," what they are saying is that the kind of love and commitment that they feel for the daughter or parent or partner is simply incompatible with making the kind of cost-benefit analysis that would justify making the life-ending decision. Others, of course, reach very different kinds of decisions.

The point that I want to make is that the question of whether a person is in fact a conscious person is always a moral question before it is a question about our justification to believe—even if it is also a question about our justification to believe. Even to raise the question of whether a person or a thing has a mind is to call one's relation to that person into question. And this is the point. For most of us, most of the time, our relations to each other simply rule out the possibility of asking the question. For the question can only be asked from a detached perspective that is incompatible with the more intimate, engaged perspective that we actually take up to each other.

It is just this nexus between morality and skepticism about the minds of others that is the enlivening theme of Ridley Scott's movie *Blade Runner*. The movie is set in a dismal future in which "replicants"—mass-produced robots—form a population of slaves. According to the dominant ideology, replicants are only machines; they have no intrinsic value; they are protected by no legal rights; they are made and destroyed according to the whims of their human owners. The twist is that some replicants have rebelled and, critically, you can't tell, by normal human interaction with one of them, whether he or she is a replicant or a real, honest-to-God person. Indeed—and this is the kicker—on the basis of introspection it is impossible for you to tell even of yourself whether or not you are a replicant. Deckard, a cop on the hunt for rebel replicants, refuses to acknowledge that the rebels are genuine conscious agents. In refusing to acknowledge their humanity, he convincingly puts his own humanity in jeopardy. For he behaves in a cruel and inhumane way. Deckard's detached judgment that the replicants are less than minded beings is deeply incompatible with the kind of relationship that he himself carries on with the "female" replicant that he befriends. That Deckard may himself be a replicant who does not know that he is one drives the point home that what is at stake here is not some kind of biological essence. What is at stake is human decency—humanity in that sense.

Man's Best Friend

In medieval Europe it was not uncommon for domestic animals—for example, pigs and asses—to be tried in criminal courts on charges such as murder and adultery. In one well-documented case, a sow was found guilty of murder in the trampling of the swineherd's boy. Despite the best efforts of her defense attorney, she was sentenced to death by hanging. Her motive was thought to have been the desire to save another pig from slaughter. Several other pigs were shown to have squealed during the general commotion surrounding the crime and were given similar sentences as accessories.

It is hard for us today even to begin to comprehend what these medieval people were thinking! Did they really believe that pigs were capable of criminal action? Or was this some sort of elaborate, indirect way of punishing the pigs' owner? Or perhaps they were trying to make a scapegoat (or rather, scapepig) of the pig. One thing is clear: this example nicely—if exaggeratedly, from a modern-day perspective—brings out the fact that domestic animals occupy a strangely complicated place in our human lives. In a way, they belong within the human sphere. We do find it very natural, for example, to think of dogs, cats, and horses as capable of complicated thought, feeling, and understanding. We have been considering the thought that our knowledge of other minds should perhaps better be understood as a kind of moral commitment to other minds. What this comes down to, in practice, is that we pursue relations and projects with other animals that close off even the possibility of calling into question their status as bearers of minds.

Let's pursue the case of dogs somewhat further. Dogs have lived in human settlements for most of recorded history. Dogs, of course, do not merely happen to be good at getting along with people: they have been bred for that purpose. It is easy to appreciate that wild dogs that displayed no apparent appreciation of human attitudes and interests would not have been taken up into human society: they would not have provoked feelings of affection, and they would not have been suitable companions. As it is, dogs are acutely sensitive to gaze, for example. Dogs make use of information about where their human companions are looking to find out where the action is, and many dogs will refuse to look a human straight in the eyes. In any case, it has been shown that domesticated animals are better able to respond to human gaze and gesture than their wild counterparts—even better than the primates.

Dogs are beloved pets in many households. But dogs are and have been, in different times and places, valued coworkers. The shepherd's sheepdog, the blind person's guide dog, the police officer's sniffer dog, the hunter's game dog—these animals enter into delicate collaborative partnerships with their human companions. What it is easy to overlook is that these relationships, even if they are hierarchical and exploitative, are richly laden with value and moral significance. The blind person does not view her guide dog as a machinelike alarm system capable of signaling the approach of obstacles, streets, damaged pavement, or other threats. The blind person relies on the dog to guide her. This relationship of guidance is one of trust and cooperation, one in which dog and human make use of a shared practice of communication and exchange. As the poet, essayist, and professional dog and horse trainer Vicki Hearne showed, training a dog to be a search-and-rescue animal—or a household protector— can never be just a matter of stimulus-and-response training. The kind of collaboration with the animal that is sought is incompatible with thinking of the animal as merely conditioned to respond to stimuli in one way or another. Training a dog is, really, educating the dog, and doing so requires that human and dog take their responsibilities to each other seriously.

This is not to insist that you cannot perfectly well treat a dog as a merely mechanistic locus of conditioned response. One can shape behaviors through reward and punishment. And, indeed, punishment and reward are likely to have a role to play in any meaningful animal/human relationship. But if one is to enter into the kind of relationship of cooperation and companionship that characterizes our actual relations with dogs, one must leave the standpoint of mechanism behind and instead view the dog as, well, a thinking being.

This attitude is all the more true of our relations to other human beings. It is possible to view a human being as a mere object, one that can be handled and manipulated to achieve whatever ends. Surely part of what explains the distinct horror of practices in Germany during the 1930s and 1940s was the willingness on the part of many to take up precisely such an objectified, mechanistic attitude to human beings, as if a human being were an item to be melted down and exploited for its trapped energy.

But if we do take up the detached, mechanistic attitude to human beings, then it is impossible for us to view them as friends or even enemies; indeed, it is impossible to think of them as genuinely subjects of experience. Once we do take up a certain kind of cohabitation with others—once we take up with them relations of friendship, marriage, collaborative working, etc.—then it becomes impossible to take seriously the thought that they lack consciousness.

What explains the extraordinary fact that we do not take skepticism about mind seriously—even though it seems, in a sense, we have good reason to—is the fact that we already occupy the nonmechanistic, engaged attitude to people, dogs, and cats. The skepticism only makes sense—it only touches us—when we are forced to step out of our normal relationships. This sense of detachment is what happens when grave injury alters those we care about or when what is at stake is our relation to other nondomestic species of animals. But once we take up the detached stance on the other, we give up any chance of understanding them.

I do not mean to suggest that a dog or a person has a mind only if we treat it as if it does—that we project mind onto things that are not genuinely conscious in themselves. Deckard's crime against the replicants, and the crimes of the National Socalists against the Jews, consisted in the refusal to accept or acknowledge the authentic humanity of the replicants and the Jews, respectively. But the fact remains: There are two fundamentally different ways of thinking about things, or two radically distinct stances we can take to things. From within one perspective, it is impossible to doubt the mind of others. From within the other, it is impossible to acknowledge it.

The Paradox of Mind and Science

We now confront a paradox. Science views its subject matter coolly, dispassionately, rationally. Science takes up the detached attitude to things. But from the detached standpoint, it turns out, it is not possible even to bring the mind of another into focus. From the detached standpoint, there is only behavior and physiology; there is no mind. So it would seem that a science of the mind is impossible. And mind itself is something paradoxical; it is a feature of our nature that cannot be made an object for natural science.

In fact, the perspective that we need, from which the meaningful, nonmechanical nature of conscious life can come into focus, is none other than the biological perspective. No living being is merely a mechanism, even though every biological system can be viewed as merely physical and so, in some suitable sense, as merely mechanical. Take, for example, a bacterium. It has size and weight and is acted on by physical forces and chemical processes. A given bacterium may move in the direction of greater intensities of sugar, owing to direct biochemical linkages between sugar-sensitive receptors and its flagella. The bacterium might seem to be geared into its environment in a machinelike way. But in fact, in thus describing the bacterium, we have already smuggled in a nonmechanical, nonphysical conception of the bacterium as, precisely, a unity, as one whose actions can be considered *as actions*, and in relation to which the question "Why?" arises. The bacterium is geared into the world not merely in the sense that the presence of sugar causes a certain bacteriumlike congeries of atoms to migrate in the direction of greater intensity of sugar; the bacterial mesh with its surroundings is of a different quality than that. The bacterium needs sugar to live and is adapted to its surroundings, and that's why it is impelled toward sugar. The bacterium is not merely a process, it is an agent, however simple; it has interests. It wants and needs sugar. Granted, the bug is not smart. That is an understatement. It doesn't understand its own reasons; there is no understanding to speak of in the vicinity. And its freedom to control the expression of its needs is no less primitive. But these are technical matters. The basic fact is that the bacterium itself only comes into focus for biology as an organism, as a living being, once we appreciate its integrity as an individual agent, as a bearer of interests and needs. With the bacterium we find a subject and an environment, an organism and a world. The animal, crucially, has a world; that is to say, it has a relationship with its surroundings.

The power of the theory of evolution by natural selection stems from the way that it naturalizes and explains this very kind of fact. To understand an animal, we must take up a perspective on its life that is at once narrative and historical and also ecological. We can ask why an organism has a certain trait, and most of the time we can provide an answer that is framed in terms of the way individuals with that trait fared better—i.e., were more likely to live to reproduce—than those without that trait. As a law, the frequency of a fitness-enhancing trait in populations always tends to rise. What this sort of approach to the animal's nature requires or allows is the acknowledgment that in nature there is more than mere mechanism. The function of a trait—which is to say, its basic significance or importance, its purpose—belongs to the ground-level explanation of why the trait even exists. But for this whole story even to make sense, we need to have the animal, the bearer of the trait, in clear focus. The animal is primary, not its traits. Nor should we think of the genes themselves as primary, as has been suggested by Richard Dawkins. The transmission of genetic information to offspring is the mechanism whereby traits are passed along; but if we want to know why certain genes are widespread in a population, we need to pay attention to the story that really matters, the story of an organism's environment-bound life history.

Physics does not catalog the existence of organisms or environments. For physics, there are only atoms and processes operating subatomically: you can't do biology from within physics. To do biology, we need the resources to take up a nonmechanistic attitude to the organism as an environmentally embedded unity. When we do that—and now we come to my critical claim—we also secure the (at least) primitive mentality of the organisms. The problem of mind is that of the problem of life. What biology brings into focus is the living being, but where we discern life, we have everything we need to discern mind.

My argument is simple: You can't have it both ways. You can't both acknowledge the existence of the organism and at the same time view it as just a locus of processes or physicochemical mechanisms. And once you see the organism as a unity, as more than just a process, you are, in effect, recognizing its primitive agency, its possession of interests, needs, and point of view. That is, you are recognizing its at least incipient mindfulness.

The problem of consciousness, then, is none other than the problem of life. What we need to understand is how life emerges in the natural world.

Mind Is Life

I said at the start of this chapter that the hardest problem would seem to be the problem of the consciousness of others. I now want to back away from that statement. What can make this

seem to be so is the fact that we think of consciousness as something that happens inside the organism, as something that happens hidden from view. All we are given are behavioral signs; we
never get at what is going on within.

In this book I am urging that we should not think of consciousness as something that goes on inside us. The mind of the
bacterium does not consist in something about the way it is internally organized. It pertains, rather, to the way it actively meshes
with its environment and gears into it. There are internal correlates of consciousness. Only creatures with the right kinds of
brains can have certain kinds of experiences, and to events in
consciousness there doubtless correspond neural events. But
there are external correlates of consciousness too. Conscious beings have worlds precisely in the sense that the world shows up
for them as laden with value: sugar! light! sex! kin! The mind of
the bacterium, such at it is, consists in its form of engagement
with and gearing into the world around it. Its mind is its life.

But the life of the bacterium is not hidden within it. The life
of the bacterium is a dynamic in which the bacterium, in its environmental situation, participates. And so it is for consciousness
more generally. To study the minds of animals, we should not
think only about the brain. To echo the language of the neuroscientist Francisco Varela and the philosopher Evan Thompson, we
need to turn our attention to the way brain, body, and world together maintain living consciousness.

Other Minds, Other Worlds

Mind is life. If we want to understand the mind of an animal, we
should look not only inward, to its physical, neurological consti-
tution; we also need to pay attention to the animal's manner of
living, to the way it is wrapped up in its place. One might have

thought that to explain the differences in the mental powers of animals, we need to appeal to ways in which they differ, one from the other—namely, to differences in their internal neurological makeup. All animals share a common external environment, after all. But in fact it is not the case that all animals have a common external environment. They may share a physical world in that, from the standpoint of physics, say, there is but one physical world. But to each different form of animal life there is a distinct, corresponding, ecological domain or habitat. All animals live in structured worlds.

Monkeys and apes, for example, live in highly structured communities: they occupy social worlds. All animals need to be able to recognize food sources, to seek out places for rest and opportunities for mating, to flee from danger and avoid predators. For monkeys living in social groups, there are social realities that make a huge difference to their lives. These factors include age, kinship, and social rank or dominance as well as patterns of alliance and cooperation. Vervet monkeys, like other primates, behave differentially toward kin. Vervet daughters help their mothers care for offspring and so establish relationships with maternal siblings. These relationships play an important role in alliances and cooperation. Vervets, for example, will be much more likely to aid another monkey in an aggressive conflict if that monkey is kin.

Do vervets have the concept of kinship? Do they understand what kinship is? In an obvious sense they do. As already noticed, monkeys treat their own kin differently than they do other group members. The phenomenon of "redirected aggression and reconciliation" is significant. Often, when two monkeys are fighting, one will attack a third monkey, a noncombatant. This redirection of aggression is kin biased; the redirection will usually be toward the opponent's kin. Moreover, reconciliation behavior, such as grooming, is similarly kin biased. One will appease an opponent by grooming the enemy's kin. There are other striking examples. There is evidence that a vervet mother will respond differently to the cries of her offspring depending on the rank of the mother of the infant with whom her own child is fighting. There is experimental evidence from captive vervets that they can consistently differentiate novel pictures of mother-offspring pairs and pairs of unrelated monkeys, even when mother-offspring pairs vary widely in their physical appearances (e.g., adult mothers and infant daughters, adult mothers and adult offspring, etc.). Similar tests show that monkeys can be trained to distinguish sibling-sibling pairs, mother-offspring pairs, and unrelated pairs. What is striking about all these cases is that they would seem to show that monkeys will not only differentiate their own kinship relations but also recognize distinct kinship relations between others in their group. It is noteworthy that one cannot explain the differentiation of kin and nonkin in terms of patterns of association among monkeys. Captive monkeys identify kin pairs, and families always act as units of alliance in intragroup competition, regardless of association rates.

There are reasons for calling into question whether vervets really know what kinship is. First of all, there is no evidence that monkeys recognize paternal relations at all. Infant vervets and fathers do not acknowledge or recognize each other. The only kinship relations among vervets that matter are bonds formed between offspring and their common mother. This finding is bizarre and striking. So whatever vervets have a concept of, it is not kinship as an abstract concept. Would it be better to think that it is a conception of *matrilineal kin group*, or something along those lines? But this brings a further problem to the fore. How could anyone be said to know what kinship (or a matrilineal kin group) is if he or she or it lacked knowledge or understanding of sexual reproduction and the biology of families? In the absence of an understanding of biology, the monkey could be said, at best, to be able to pick out the matrilineal group, but that would not show that it understands, in any interesting sense of the word, what such a matrilineal group is. And there are other gaps in their putative understanding. For example, there is no evidence that monkeys can recognize relations of kinship as holding between members of other species.

But this skepticism about monkey minds misses the point. To describe the world of the monkey, you must describe a world in which relationships are structured, as we would say, along dimensions of dominance and kinship; and clearly, the monkey is right at home in this world. Granted, we may be uncertain whether we should say that the monkey has any grasp on what *we* call kinship or that it has a confident grasp on a notion more primitive than kinship. But this indeterminacy notwithstanding, we must appreciate that the monkey's very mode of being is one in which kinship relations matter, and this mode of monkey being amounts to a form of substantial cognitive achievement.

Conclusion: Understanding Ourselves

Life is the lower boundary of consciousness. I do not know where we will find its upper limit. I don't rule out the possibility of artificial robot consciousness. But I would not be surprised if the only route to artificial consciousness is through artificial life.

The question of consciousness arises for living beings and it arises for them because living beings exhibit at least primitive agency. To study mind, as with life itself, we need to keep the whole organism in its natural environmental setting in focus. Neuroscience matters, as does chemistry and physics. But from these lower-level or internal perspectives, our subject matter loses resolution for us.

What I am saying is that the question of consciousness arises for living beings. To answer the question—about this or that organism—we need to look to the details on a case-by-case basis. Vervet monkeys have minds appropriate to their lives; they are not mere machines. If I am on the right track, then, however far-fetched, something of this sort should be said about the bacterium itself. It is a primitive agent, which is to say, it is a primitive subject.

The link between life and consciousness is critical. Part of what makes it so hard to judge whether a person in a persistent vegetative state has experience is that her life has been utterly disrupted; in a way, her life itself is called into question for us. Matters are different when we ask if, say, a lobster feels our touch. The lobster's life is questionable for us, not because it is disturbed, but because it is so alien. In neither case, though, is our problem that we cannot pull back the curtain and look inside. The organism's life is not inside.

In the next chapter I turn to the brain and to the task of understanding the brain's role in explaining animal consciousness. I show that the perspective developed here provides a new setting in which to understand and explain consciousness.

3

THE DYNAMICS
OF CONSCIOUSNESS

Where do we find ourselves? In a series of which we do not know the extremes, and believe that it has none. We wake and find ourselves on a stair; there are stairs below us, which we seem to have ascended; there are stairs above us, many a one, which go upward and out of sight. But the Genius which, according to the old belief, stands at the door by which we enter, and gives us the lethe to drink, that we may tell no tales, mixed the cup too strongly, and we cannot shake off the lethargy now at noonday.

—Ralph Waldo Emerson

How does consciousness arise in the brain? In this chapter I offer evidence that it does not. We can explain how the brain's activity gives rise to consciousness only when we appreciate that what matters for consciousness is not the neural activity as such but neural activity as embedded in an animal's larger action and interaction with the world around it. Which is another way of saying that it isn't the neural activity on its own that fixes consciousness. I propose that the brain's job is that of facilitating a dynamic pattern of interaction among brain, body, and world. Experience is enacted by conscious beings with the help of the world.

Magical Membranes

What features of cells in the brain enable us to see and feel and wonder? This is a trick question. If there is one thing we now know, it is that the character of human experience is not fixed by the properties of individual neurons. The neuron is just the wrong unit of analysis. Brain cells are pretty much all the same. They have the same basic plan—cell bodies innervated by bushy dendrites and activating voltage transmissions along wirelike axons—and they behave in roughly the same ways: they participate in patterns of electrochemical activation.

You can no more explain mind in terms of the cell than you can explain dance in terms of the muscle. If the character of our mental lives depends on what's going on in the brain—and it does—then we need to turn our attention away from individual neurons. It is now accepted that if we are to have any chance of understanding the brain basis of consciousness, we need to widen our gaze to encompass large-scale populations of neurons and their dynamic activity over time.

But why stop there? It isn't as though brain function is transparent when we look to the dynamics of large-scale assemblies of cells. If you were to land in the brain like a microscopic alien, you would be unable to tell, by inspecting the local neural fireworks, whether there was even experience going on, let alone whether that experience was visual, say. Perhaps what accounts for this explanatory opacity of the neural systems is that even large-scale neural dynamics do not yet provide the right level of analysis at which to make sense of animal consciousness. Just as the fact that we cannot understand phenomena of consciousness in terms of the individual cell leads us to consider the causal powers of populations of cells, so the limits of what we can understand in terms of populations lead us to expand our conception and think of neural systems as elements of a larger system that includes the rest of the animal's body and also its situation in and interaction with the environment. Perhaps the proper scale at which to make sense of neural function—that is, of the contribution that brain makes to mind—is that of the living, environmentally situated animal itself.

If this seems like an unlikely proposal, it may be because tradition teaches that the skull is the boundary marking off what is inside from what is outside. And, crucially, we are inside: mind depends only on what happens within us. According to the traditional perspective, the role of the "external world" in consciousness is merely to provide outer irritation or mere peripheral impingement. But why should we think that the boundary formed by the limits of the brain is special when compared with other boundaries that can be drawn inside the brain (e.g., between individual cells, or populations of cells, or areas of the brain)? As Susan Hurley liked to say, the skull is not a *magical membrane*; why not take seriously the possibility that the causal processes that matter for consciousness are themselves *boundary crossing* and, therefore, world involving?

My claim is that we need to take this possibility seriously if we want to understand consciousness. In this chapter I'll begin to show why.

No Man Is an Island

The infant brain is changeable and plastic. Sensory stimulation produces the very connectedness and function that in turn make normal consciousness possible. For this reason, sensory deprivation may produce permanent damage. Hubel and Wiesel showed this by rearing cats in the dark; what they demonstrated was that cats deprived of sight during a critical period in infancy would never be able to see. The neonatal mammal, we learn, is plastic and open; in a very real sense the environment itself produces in us the conditions needed to experience that environment. They also showed that there are limits to the brain's plasticity.

Of special importance to the child's neurological development is its relationship to other people. Consider a question explored by Bruce Wexler in an excellent recent book on this topic: Why do mammals suckle at the breast? To get food, to be sure, but also to get touch—that is, to get the stimulation that is itself necessary sustenance for the developing brain. Kenneth Kaye has shown that feeding may play a foundational role in the child's mental development. All human mothers that have been studied (and only human mothers, it turns out) spontaneously jiggle their suckling young when they pause in their drinking (this goes for bottle-fed babies too); babies in turn spontaneously wait until mother has stopped jiggling before resuming their sucking. Kaye suggests that this is a primitive form of taking turns. It is not difficult to see this as a kind of proto-conversation or, at least, a necessary antecedent to that uniquely human form of communication.

Mothers or other primary caretakers do not merely take care of their young. A two-way exchange arises between child and mother that provides the setting within which the child develops both physiologically and psychologically. A child learns to restore its own calm or comfort by being calmed or comforted by the mother. The child's basic physiological processes—burping, for example—are facilitated for the baby by the mother. The caretaker manipulates the child's posture, raising her into a sitting position, now setting her down in a prone position, either to arouse the child or prepare her for sleep. The caretaker directs the child's attention to this or that and manipulates things for the child. The mother's attentiveness to the needs of the child teaches the child to learn to manage her own needs. In a very real sense, the baby-caretaker "dyad" is a unity from which the child only gradually emerges as an individual. We can speak of attachment here, but I prefer to speak of oneness. Our separation from our mother-figure is, in some respects—for most of us, anyway—only partial; in any event, there is for us no such thing as complete detachment from the community of others and from the larger environmental structures and situations—lights, sounds, odors, the ground, the air, technology—up against which we first become ourselves.

Maturation is not so much a process of self-individuation and detachment as it is one of growing comfortably into one's environmental situation. We grow apart, but we attach to the world without. We integrate. In learning to walk or mastering language, in developing friendships, in acquiring an occupation, in learning to navigate and use technology, we root ourselves in the practical environment. This is one reason, certainly, why radical changes to one's environment, especially occurring later in one's life—for example, in the course of migrating from one country to another, upon the loss of a spouse, during a period of rapid technological change—are enormous, maybe even devastating personal challenges. The loss of a feature of the environment with which one's daily activities are intimately interwoven is the loss of a part of oneself. I return to this theme in the next chapter.

The point is not that you can't teach an old dog new tricks: sometimes you can. The point really is that to do so, you need to make the dog new again. Someone once told me that it is good to change jobs every seven years or so. It keeps you young. One explanation for this might be that changes force you to renew yourself by developing anew in relation to new external structures, new habits, new modes of involvement with the world around you. Another effect of this kind of disruption is that time, in an interesting, felt way, slows down. When life is routinized, days and weeks and months blend into one another: each day is like the next; the days form an arc as one's life project unfolds.

But when the routine is disrupted—when you move, or even when you travel—days acquire a distinct specialness. A week getting settled in a new city can seem like a lifetime! It is a tantalizing trade-off. One gives up comfort and, in a way, productivity; in return, one gets time and youth!

We see something like this very dynamic in the development of a young person. A summer can seem like a magical eternity to a child. Think of the play and sweat and swimming and books and people and shopping and long evenings! But getting older requires one to channel one's day into a structure of meaning, into a plan; under the shadow cast by one's projects, life acquires an organization that in a way removes the surprise. To grow old is to give up surprises; to insist on surprises is, in a way, to stay young. This may be one reason why we have kids.

The cost of changing one's life situation on a regular basis is high. We seem to get a bit tired as we grow old, and it takes a lot of energy to refit oneself to a new world. On the other hand, there are achievements that can be had only by one who is comfortably settled into a way of life. The positive pleasures of domestic life, for example, or the skills that come from practicing a craft or trade or profession with perfection and mastery, will probably be sacrificed if one moves around too much.

Neural plasticity, properly understood, teaches us that the brain can never be the whole story about our mental development. As Wexler observes, no other animal can develop the linguistic capacities of even a person whose "language centers" have been surgically removed from his or her brain in infancy. Our linguistic capacity, it follows, is not a product of a particular neural structure. Language is a shared cultural practice that can only be learned by a person who is one among many in a special kind of cultural ecosystem. Neural plasticity can also teach us a good deal about consciousness, as I will now show.

Neural Plasticity and Consciousness

What is it about the character of brain activity that enables us to have visual experience as opposed to some other kind of experience (auditory or olfactory, for example) or even no experience at all? This is a question about consciousness, about the distinct qualitative character of our experiences. Specifically, it is a question about the neural basis of the qualitative character of our conscious episodes. What is it about the distinctive neuronal snap, crackle, and pop that makes the resulting experience have one specific kind of felt character rather than another?

Scientists have been unable to answer this question. Until now, they have not been able to bridge what is sometimes called "the explanatory gap" between neural states on the one hand and conscious experience on the other. From my perspective, this is not surprising. The reason we cannot explain the quality of experience in terms of the intrinsic nature of the brain's action is that, in fact, there is nothing distinctively visual about the brain's action.

Let me explain this. Consider a group of startling and highly instructive studies on ferrets that were done by Mriganka Sur and his colleagues at MIT. Sur and his team operated on newborn ferrets, in effect wiring up the eyes to the parts of the brain normally used for hearing. What they actually did was alter each ferret so that cells in the eye that would normally sprout connections into the visual areas of the brain (the visual thalamus, the visual cortex) instead grew into the areas of the brain normally dedicated to hearing. Because ferrets are born in a neurologically very immature state, they are ideal for this sort of intervention.

You might have thought that rewiring ferrets in this way would cause them to hear with their eyes. After all, the eyes were wired up to the hearing parts of the brain. Instead, it enabled them to see with their auditory brains. This is a remarkable finding. It shows that the link between brain areas and conscious experience (e.g., the link between the auditory cortex and auditory experiences and that between the visual cortex and visual experiences) is malleable. What Sur did, in effect, was bring about the undoing of the normal correlation between neural activity in a given brain area and visual experience. Normally, neural activity in the visual cortex gives rise to the experience of seeing. But in these altered ferrets, the animals see when an entirely different part of their brain is activated. By modifying the normal relation between the eyes (or retinas) and the brain, Sur brought about a remapping of experiences and the brain. (In fact, he rewired only one hemisphere, so the ferrets were able to hear perfectly well using the non-rewired auditory cortex.)

The fact that it is possible in this way to vary consciousness in relation to its neural underpinnings teaches that there isn't anything special about the cells in the so-called visual cortex that makes them visual. Cells in the auditory cortex can be visual just as well. There is no necessary connection between the character of experience and the behavior of certain cells.

And this finding in turn means that if we want to understand why certain cells or certain brain areas are participating in seeing and not hearing, or in hearing and not seeing, we need to look beyond the immediate neural activity itself. The character of conscious experience can vary even though the neural activity underpinning it does not change. This is the basic lesson of Sur's studies. It follows, then, that what determines and controls the character of conscious experience is not the associated neural activity.

Bridging the Gap

We can begin to get a handle on the relation between consciousness and the brain by looking at situations in which the normal correlation between neural activity, on the one hand, and experience of one or another qualitative character, on the other, is altered. That is, we can use the fact of the brain's plasticity as a probe with which to try and figure out why neural activity is bound up with experience in the ways that it is. This is the strategy Susan Hurley and I pursued together over a number of years, until her death in the summer of 2007.

Sur's ferrets provide one such example. It can be contrasted with a different sort of case, the well-known but nonetheless bizarre phantom limb phenomenon. Sometimes, when an amputee is touched on the face, she will report feeling as if she is touched on the now missing hand. Why? Hand and face areas are next to each other in the cortex. After amputation, the hand area is idle, disused. It would appear that the neighboring face area invades the hand cortex or somehow gets entangled with it. As a result of this entanglement, touching the face now produces two distinct cortical effects. First, it produces an activation in the face cortex corresponding to the feeling of being stroked on the face. Second, it produces an activation in the hand cortex corresponding to the feeling of being touched on the now absent hand. Touching the face produces a feeling of being touched on the hand for the same reason that pressing the doorbell will turn on the lights if the doorbell has been wired up to the light switch. Touching the face simply brings about those neural effects that would have been brought about by the touching of the hand.

The contrast between Sur's ferrets and the case of the phantom limb is striking. Sur's ferrets don't hear with their eyes, they see with their auditory brains: the auditory cortex changes its function for consciousness as a result of receiving stimulation from the eyes. But in the case of "referred sensation" to a phantom limb, things are just the other way around. The phantom limb patient, crazy as this may sound, feels on his hand with his face (instead of feeling his face in his cortical hand area). Why?

Because the face is wired up to the hand area of the cortex. Activation of the hand area continues to give rise to the feeling of being touched on the hand, even though the activation stems from touching the face, not the hand.

Why does a change in the source of stimulation of a cortical area sometimes bring about changes in the character of a resulting experience, as in the case of Sur's ferrets, and sometimes fail to do so, as in the case of the phantom limb?

At the end of this book I give references to an essay in which Hurley and I tried to supply an answer to this in full detail. But the short answer can be given here. In a way, I have already given it. What explains the consequences for consciousness of this sort of rewiring is not the intrinsic character of the neurophysiological changes themselves; it is, rather, the larger setting or context in which these neurophysiological changes occur.

Now the question we need to pose is, What is the relevant larger context in terms of which we can hope to understand the effects of neural rewiring on conscious experience? Put more simply, what are the contextual factors governing the character of experience and its relation to neural activity? I will now show you that the relevant context must go beyond that of the brain itself to include the animal's active relation to its surroundings.

Sensory Substitution

We have already seen that rewiring is not sufficient for a change in the quality of the associated experience. That's what the phantom limb case proves. It turns out that rewiring isn't necessary for changes in consciousness, either.

In the late 1960s, the engineer and physiologist Paul Bach-y-Rita developed a device to enable blind people to see. He continued to work on this project until his death two years ago. Remarkably, Bach-y-Rita actually succeeded. For a variety of reasons, his device has been considered impracticable; it is cumbersome and ultimately not ready for daily use. But from a theoretical standpoint, the work is of great consequence. His starting point was the belief that the eyes are a channel for getting information to the nervous system; therefore, it ought to be possible to provide the same visual information to the brain through a different channel. He set about devising a way to do this. Here's what he came up with: He wired a camera to an array of vibrators that he placed on the thigh or abdomen of subjects. The wiring was such that visual information presented to the camera produced a range of tactile stimuli on the subject's skin. What he found was that when the camera was mounted on the head or shoulder of the person, visual information presented to the camera that in turn produced tactile sensations on the body enabled the person to make judgments about the size, shape, and number of objects placed on the other side of the room. By deploying the substitution system, the blind person was able to reach out and pick up objects, and even swat at a ball successfully with a Ping-Pong paddle. This is astounding. In effect, blind subjects using a tactile-visual substitution system can see! Somehow, for a person who has a few hours to get used to the apparatus, a series of tactile sensations on the leg or stomach add up to a way of seeing.

Now, I want to emphasize that tactile-visual sensory substitution is a full-fledged, bona fide example of the kind of transformation in perceptual consciousness that we saw with Sur's ferrets. Stimulation of the skin gives rise to neural activity in touch areas of the brain (the so-called somatosensory cortex). But for a person who has adapted to the sensory substitution system, activation in somatosensory touch areas gives rise not to the experience of being touched (or at least not only to the feeling of being touched) but to a visual experience of the scene in front of him. Touch areas of the brain are changing their function for consciousness as the ferret's auditory cortex changed its function for consciousness. But in sensory substitution, in contrast with Sur's ferrets, we cannot hope to explain why the cortex changes its function for consciousness by supposing that it is receiving stimulation in a novel or surprising way—for the simple reason that it is *not* receiving stimulation in a novel or surprising way. Vibrations on the skin are activating the somatosensory cortex in the good old-fashioned way. Nor is it very plausible that the somatosensory cortex has undergone any particular kind of neurophysiological reorganization. After all, Bach-y-Rita used full-grown and therefore relatively nonplastic adults as his subjects. Moreover, he found that people adapted to his sensory substitution not in weeks or days but in hours and minutes. That's just not enough time for any significant internal rewiring to occur.

Bach-y-Rita's sensory substitution system is perceptual plasticity without neural plasticity. What better reason could there be to acknowledge that we need to look beyond the brain if we want to get a handle on what is bringing about the dramatic changes in the character of experience that we witness? But where should we look? What explains the change in the qualitative character of experiences associated with the somatosensory cortex, if in fact there is no rewiring, or corresponding change in neurophysiology?

Looking Beyond the Brain

We are so much in the grip of the idea that our experience is governed by neural events inside us that we miss what is really the natural, obvious explanation of the experiential changes we see in tactile-visual sensory substitution. Bach-y-Rita himself thought that the purpose of his system was to deliver visual information to the brain in a novel manner. But ask yourself: What does Bach-y-Rita's sensory substitution system actually do? At the most basic level, it sets up a relation between the perceiver and objects in the scene around him where there was no relation before. When you are outfitted with the system, the stimulation of your skin is now affected, in a novel but entirely systematic way, by changes in your spatial relationship to objects. In effect, the tactile-vision substitution system has established a new way of being connected to the environment.

This is the key to our puzzle. What governs the character of our experience—what makes experience the kind of experience it is—is not the neural activity in our brains on its own; it is, rather, our ongoing dynamic relation to objects, a relation that, as in this case, clearly depends on our neural responsiveness to changes in our relation to things. It is there, in this extended, sensorimotor involvement with the world, that we find the resources to explain why we see when we use the tactile-visual sensory substitution system. My proposal, specifically, is this: We *see* with Bach-y-Rita's system because the relationship that system sets up and maintains between the perceiver and the object is, in ways that can be made precise, the sort of relation that we bear to things when we see them. What causes the effects for consciousness of neural activity in the touch-dedicated parts of the brain to change? Answer: the world and our relation to it.

Action in Perception

Traditional approaches to vision have tended to suppose that vision happens in us. It is a phenomenon of the retina and structures in the brain. I'll discuss this idea, and some problems it gives rise to, in more detail later, in Chapters 6 and 7. For now, I want to point out what ought to be entirely obvious anyway, namely, that seeing is, in many ways, a bodily activity. Seeing involves moving the eyes and head and body. More important, movements of your eyes or your head or your body actively produce changes in sensory stimulation to your eyes. Or, put differently, how things look depends, in subtle and fine-grained ways, on what you do. Approach an object and it looms in your visual field. Now turn away: it leaves your field of view. Now shut your eyes: it is gone. Walk around the object and its profile changes. In these and many other ways, there are patterns of dependence between simple sensory stimulation on the one hand and your own bodily movement on the other. It should be clear that a central task for any perceiving organism is to master these dynamic patterns of sensory stimulation and movement.

This suggests a whole new way of thinking about what perception is. Indeed, Kevin O'Regan and I have developed just such an approach. According to this sensorimotor, enactive, or actionist approach, seeing is not something that happens in us. It is not something that happens to us or in our brains. It is something we do. It is an activity of exploring the world making use of our practical familiarity with the ways in which our own movement drives and modulates our sensory encounter with the world. Seeing is a kind of skillful activity.

What makes vision visual is that it is an activity of exploring the environment making use of an understanding of the specific family of ways in which movement produces sensory change, namely, those ways that depend, centrally, on the eye. Blinking, turning the eye or head, and moving in relation to objects bring about characteristically eye-based sensory events. What is remarkable is that the world can show up for visual consciousness—objects can show up with all manner of spatial and visible properties—thanks to our fluent appreciation of the ways that eye-related visual sensory stimulation depends on our own movements.

Other sensory modalities, such as hearing and touching, are no less ways of exploring the environment, but the way the world shows up for auditory or tactile experience depends on entirely different patterns of sensorimotor interdependence. Consider that there is no simple tactile sensation of squareness. To perceive the squareness of something by touch is for one's sensory interactions with the thing to be structured in a specific range of ways—that is, for one's movements to be obstructed or guided in precise ways. Remarkably, one can have a sense of the squareness of something even when one is only touching one corner. The squareness shows up for touch even in this case because one understands, in a practical sort of way, the kinds of movements that would be allowed by the object's contours. Likewise, when I see a house, I can have a visual sense of the presence of the whole house even though, in fact, all I can take in from where I am standing is, say, the house's façade. The rest of the house can be present for me because I understand, in an implicit, practical way, that my relation to the house as a whole is mediated by a distinctive repertoire of exploratory skills.

In this approach, sensory modalities are really styles of exploration of the world, and they differ from each other the way that musicians can differ from each other in their styles—that is, in the battery of movements and expectations and skills that they deploy when playing their instruments.

Back to Sensory Substitution

From this standpoint, we can appreciate what it is about Bach-y-Rita's sensory substitution system that makes it visual. It is visual because it is a way of exploring the world that depends on the exercise of what happens to be visual sensorimotor understanding. That is to say, the ways in which sensory stimulation depends on movement in tactile-visual sensory substitution is similar to the ways in which it depends on movement during vision. They share a style. For example, just as with normal vision, in tactile-visual sensory substitution, things get bigger as you approach them. And turning your body away from them leaves them out of view.

One can grant this without also holding that tactile-visual sensory substitution is exactly like vision. The action-oriented standpoint I have been laying out here can accommodate this point. Indeed, it predicts it. After all, if perception depends on sensorimotor skill, then it also depends on the precise character of our bodies, for our skills are wrapped up with the way we are built. There are bound to be important differences between tactile-visual sensory substitution and vision just because of the very different manner in which perceptual activity gets embodied in these two ways of exploring the world. It is only when we step back and look at tactile-visual sensory substitution and vision at a more abstract level of characterization that we can see them as essentially isomorphic in their sensorimotor structure. One thing is clear and worth noticing, even if we want to hold the line and insist that there is nothing visual about tactile-visual sensory substitution: it is not tactile, either. That is, there is nothing even remotely tactile about this way of perceiving. In touch we find out about things around us by coming into contact with them, but tactile-visual sensory substitution enables us to tell how things stand around us and at a distance, in precisely the way that vision does.

You may object that when you explore the world with this sensory substitution device, you will experience vibrations on your skin. There is evidence that one can, with effort, turn one's attention to these sensations. But this does not show that tactile-visual sensory substitution is just a matter of buzzing on your skin. For one thing, it requires a deliberate, conscious effort to turn your attention away from the world you are interested in and focus instead on this buzzing, just as it would require effort to turn your attention from what you see and focus on the feel of the arms of the eyeglasses on your ears. But in any case, the buzzing is at most a causal by-product of the functioning of the system. To confuse these felt vibrations with the modality of the experience is like thinking that seeing with sunglasses is really tactile because it is possible to direct one's attention to the feel of the cold metal of the frames on your skin. The feel of the frames, like the buzzing on your thigh, are accompaniments of the perceptual activity and not constituents of it.

The Upshot

The central claim of the theory of perceptual consciousness that I am laying out here is that it is not the intrinsic character of sensory stimulation that fixes the character of experience; rather, it is the way sensory stimulation varies as a function of movement in relation to the environment that does the important work. And this is precisely what we see when we look at tactile-visual sensory substitution. The visual (or quasi-visual) character of the sensory substitution system is not fixed by the nature of the neural activity in the somatosensory cortex; rather, it is fixed by the ways in which that activity varies as a function of movement. And crucially, the *way* that activity varies as a function of movement is precisely the visual way. To be the visual cortex—that is, to play that functional role—a cortical region has to occupy a position in a very specific dynamic sensorimotor context.

Now, from this standpoint we can also understand precisely what it is about the rewiring of a ferret's eyes to the auditory cortex that induces the change in the qualitative function of the ferret cortex. The surgical intervention has precisely the effect of allowing for the establishment of a visual sensorimotor dynamic structure: what makes the auditory cortex the visual cortex for the ferret is the fact that it is drafted into that visual sensorimotor dynamic structure. Indeed, in the phantom limb scenario, it is precisely because the hand cortex is not drafted successfully into a new structure that neural activity in the hand cortex effectively dangles. The phantom limb is the consequence of this failure of dynamic integration of neural activity in the hand cortex.

Brain and World

What explains the plasticity of consciousness that we witness in the ferret case—and also in tactile-visual sensory substitution—is not something that can be understood just in terms of the relation between neural activity and the sensory periphery of the nervous system (e.g., the retina or cochlea or the receptors in the skin). To understand the sources of experience, we need to see those neural processes in the context of the conscious being's active relation to the world around it. We need to take into our purview dynamic relationships that cross the not-so-magical membrane of the skull. Consciousness of the world around us is something that we do: we enact it, with the world's help, in our dynamic living activities. It is not something that happens in us.

The beauty of this framework for thinking about the brain and human experience is that it enables us to appreciate why the brain is vitally necessary for human experience without treating the brain as if it possessed magical powers. The brain does not generate consciousness the way a stove generates heat. A better comparison would be with a musical instrument. Instruments don't make music or generate sounds on their own. They enable people to make music or produce sounds. Crick's idea that you are your brain—or, in more basic terms, the idea that consciousness is a phenomenon of the brain, the way digestion is a phenomenon of the stomach—is as fantastic as the idea of a self-playing orchestra.

The brain has a job to do; I have tried to convince you that a careful examination of the way experience and the brain's activity depend on each other makes plausible the idea that the brain's job is, in effect, to coordinate our dealings with the environment. It is thus only in the context of an animal's embodied existence, situated in an environment, dynamically interacting with objects and situations, that the function of the brain can be understood.

This is a dramatic outcome of our investigation. For one thing, it means that the world itself can be described as belonging to the very machinery of our own consciousness. This isn't poetry; this is a well-supported empirical hypothesis. Perceptual consciousness, at least, is a kind of skillful adjustment to objects (and the environment). Seeing is a style of skillful interaction with the things we see. We couldn't do that if we had no brain, but we couldn't do it if there were no objects, either. The same point can be made about the body (that is, the rest of the body that is not our brain): the body gives structure and shape to the kinds of relations we can have to the world around us; the world shows up for us thanks to our bodily ability to coordinate our relation to it. And of course the brain, too, is a necessary element in the story.

Conclusion: The Machinery of Mind Is Extended

We began by asking what features of individual cells explain the qualitative character of human experience. We conclude by appreciating that we are looking for consciousness in the wrong place if we look for it in the brain. We need to widen our conception of the machinery of consciousness beyond the brain to include not only the brain but also our active lives in the context of our worlds. This is what the biology of consciousness now teaches. In the next chapter I begin to explore some of the implications of this new way of thinking about our conscious being.

4

WIDE MINDS

We know better now. We know that that life of man whose
unfolding furnishes psychology its material is the most difficult and complicated subject which man can investigate. We
have some consciousness of its ramifications and of its connections. We see that man is somewhat more than a neatly
dovetailed psychical machine who may be taken as an isolated individual, laid on the dissecting table of analysis and
duly anatomized. We know that his life is bound up with the
life of society, of the nation in the ethos and nomos; we
know that he is closely connected with all the past by the
lines of education, tradition, and heredity; we know that
man is indeed the microcosm who has gathered into himself
the riches of the world, both of space and of time, the world
physical and the world psychical.   —John Dewey

In the previous chapter I offered evidence that the brain gives
rise to consciousness by enabling an exchange between the person or animal and the world. What emerges from this discussion
is a new conception of ourselves as expanded, extended, and dynamic. In this chapter I place this discovery in a larger context.
Our bodies and our minds are active. By changing the shape of
our activity, we can change our own shape, body, and mind. Language, tools, and collective practices make us what we are. Where
do you stop, and where does the rest of the world begin? There

is no reason to suppose that the critical boundary is found in our brains or our skin.

Where Do We Find Ourselves?

We now think of economies as globalized, corporations as internationalized, information networks as distributed. We ourselves are also dynamically distributed, boundary crossing, offloaded, and environmentally situated, by our very nature. What explains our inability until now to understand consciousness is that we've been searching for it in the wrong place.

I remember vividly one day in the 1970s—I must have been eleven or twelve—when my father addressed me sharply on the street in front of our building in New York: "Stop acting that way immediately. You're behaving like . . ." He paused to find the right word. "You're behaving like an American!" I don't recall what I was doing. Perhaps I was singing aloud, or doing a little dance step, or in some other way carrying on in public. My father was an immigrant who'd arrived in New York at the end of 1949, having made it through the clutches of the Nazis and the Soviets; he was grateful to America—convinced he would not have survived but for their intervention in the war—and glad to enjoy the freedom and anonymity of New York City. But a part of him, at least sometimes—so I was startled to discover—was anything but at home here. America, for my father, was brash, unmannered, loud, superficial, and above all foreign. Although he loved me, at that moment, at least, he was revolted—I don't think that is too strong a word—by his very own son for being what he—that is, what I—could not help being: a native of this new place.

I've since learned that this is a very common bind for immigrants. It is usually hardship that forces them to seek a home in a strange land. But what struck me—and I think I appreciated this even as a child, in the face of my father's displeasure—was the thought that he was divided against himself. The revulsion he felt toward me and toward the place where he found himself showed just how displaced and indeed, in a way, disfigured he was.

This is a book about consciousness, about the human mind and the project of understanding it as belonging to our biological natures. It is not a memoir or a tale of my father's immigration. I mention this personal anecdote because it underscores an important idea—and also because my own preoccupation with this problem may have had its beginnings in this early experience. What my father's plight illustrates is that, at a very basic level, we are *involved*—that is to say, tangled up—with the places we find ourselves. We are *of* them. A person is not a self-contained module or autonomous whole. We are not like the berry that can be easily plucked, but rather like the plant itself, rooted in the earth and enmeshed in the brambles. When we transplant ourselves as immigrants get transplanted, when we move from one town to another or one country to another, we suffer injury, however subtly or grotesquely or even painlessly, and so we are altered. This should not come as a surprise. Our life is a flow of activity, and it depends on our possession of habits and skills and practical knowledge whose very actuality in turn implicates our particular niches. No matter how good you are at breathing, you can't breathe underwater, just as you can't swim where there is no water. And no matter how charming you may be, how wonderful a raconteur, if you find yourself in a strange land where a strange language is spoken, you can't tell a good story— that is, you can't be what you are. You yourself are changed.

Where do you stop, and where does the rest of the world begin? The blithe confidence of the neuroscientist that the brain is the seat of consciousness amounts to an unearned conviction that we can draw the boundary between ourselves and the rest of the world at the skull. For some purposes that may be a good way to go. If you want to know how many people are in attendance at the ball game, don't count the number of arms: count heads. But counting depends on settled ways of individuating that which we wish to count. It is a highly purpose-relative activity. Are the dandelions in the meadow one plant, joined as they are through a common root system, or many? Is the Macintosh OS one program, the operating system, or is it many: a mail program, a calendar program, etc.? Is Williams-Sonoma one company or many? How we draw these lines usually depends on what we are interested in. Are we programmers or potential investors? Do we work for the antitrust division of the Justice Department? It also depends on what we want to accomplish. One of the central claims of this book is that if we seek to understand human or animal consciousness, then we ought to focus not on the brain alone but on the brain in context—that is, on the brain in the natural setting of the active life of the person or animal. For what we bearers of consciousness are—as the example of my father serves to illustrate—depends on where we are and what we can do.

Magical Boundaries and the Rubber-Hand Illusion

Where do you stop, and where does the rest of the world begin? One extreme view would have it that you are your brain: you stop at its limits. The skull, roughly, is the boundary of yourself. You may think that the pain of stubbing your toe is in your toe and that the feeling of the glass is in your hand, but you are mistaken. The feeling itself is not in your hand or toe; it is *in your brain*. Granted, it is the action of the cup on the sensory fibers in your hand, or the activation of nerve endings in your toe, that causes the neural activation in your head in which your feeling consists. But you'd have the feeling even if there were no cup or no foot, just so long as the right pattern of activation was brought about. The real sensation is in your head, not in your body. Something vaguely like this idea may inform people who say that the brain is the most important erogenous zone.

This claim is yet another version of the prevalent neuroscientific dogma that you are your brain and that all the rest—the sense of our emplacement in a world that is meaningful and populated by others—is a myth promulgated for us by our brains. However intoxicating it might be to think that science has this to teach us—"we live behind a veil of illusion"—there is no reason to be convinced. The established facts are only these: sensation requires the action of the nervous system; there is no human or animal life without a nervous system. But from this it does not follow either that the nervous system is alone sufficient for sensation or that our selves are confined to our brains and nerve tissue.

The rubber-hand illusion, as it is sometimes called, provides a lovely piece of support for what I am claiming. This demonstration, first performed by Matthew Botvinick and Jonathan Cohen and reported in the journal *Nature* in 1998—in an article entitled "Rubber Hand Feels Touch That Eyes See"—is a stunning illustration of the fact that the sense of where we are is shaped dynamically by our interaction with the environment in multiple sensory modalities.

The demonstration went like this (I simplify only slightly): You are asked to sit at a table. Your right hand is on your lap and is concealed from your view by the table. Across the table rests a rubber hand, the sort of thing you might find in a shop selling scary toys for Halloween. You watch as a person gently taps and strokes the rubber hand with a delicate paintbrush. Tap tap, stroke stroke, tap stroke, stroke stroke stroke tap. In perfect synchrony, an experimenter is tapping your *actual* right hand out of view underneath the table. Now something remarkable happens. You have the very distinct feeling that you are being touched on the rubber hand: that the feeling of being touched that is, in fact, occurring on your own, connected right hand under the table is taking place on the rubber hand across the table! As the article's title suggests, you feel the touching of the rubber hand that the eyes see. If you are asked to point with your left hand, which is also under the table, to the place where you feel yourself being touched, you will point (roughly) in the direction of the rubber hand.

This is an example of a very striking and prevalent phenomenon: the power of what we see to influence our nonvisual sensory experience. This is known in psychology as visual capture. It is the phenomenon that underlies ventriloquism. What causes you to hear the words coming from the mouth of the dummy is the fact that you see the dummy's mouth opening and closing in sync with the words themselves. You hear what you see. This is a robust effect; we experience it when we go to the movies, where the actor imaged on the screen seems to be producing the sounds. In fact, the sounds emanate from speakers that are located elsewhere in the room. In one important study, speakers play two distinct unrelated streams of speech, creating a jumble of speech noise. You can't understand either stream until you are given visual cues in the form of video images of the faces doing the talking. The spatial distinctness of the visually apparent sources of the sounds enables you to discern the distinct streams of speech when you were unable to do this otherwise.

In fact, visual capture—the powerful influence of vision on other sensory modalities—is an important element in normal, true speech perception. We unconsciously read lips when we are engaged in conversation, and what we hear—what sounds we take in—depends critically on what we see. To different speech sounds there correspond distinct patterns of lip movement and mouth shaping. Part of what enables us to succeed in hearing the speech sounds correctly is that we see what sounds are being produced. It's important to realize that it is difficult to hear speech sounds. The acoustic stimulus is liable to be heard in different ways. We experience this when we try to spell an unfamiliar word or an unusual name on the telephone. If you want to be sure that the airline people get your name right, you had better spell it out using conventional ways of naming the letters (e.g., "Romeo" for *r*, "Alpha" for *a*, etc.). Otherwise, they just won't be able to understand you.

A demonstration of the robustness of the influence of seeing on hearing is the McGurk effect, named after the developmental psychologist Harry McGurk. You hear a recording of someone saying "ba," but you see a synchronized video image of someone saying "ga." What you experience is the video head producing the sound "da." Whatever the details of the explanation of this phenomenon, the basic idea would seem to be clear: we use information in the acoustic stimulus and information in what we see—how the mouth is moving—in order to achieve a perception of a speech sound itself. If these sources of information are not consistent with each other, our ability to perceive correctly what is said breaks down.

It is probable that our ability to hear words also depends, to some degree, on our knowledge of what is being talked about and our expectations of what will come next. The linguist Geoffrey Pullum once offered a nice example of this in conversation. You say to someone, "Here is a hat, here is a scarf, here is a dlove." Invariably he or she will actually hear that last word as "glove." "Hat" and "scarf" prime the listener for another article of winter clothing, such as a glove; moreover, the "dl" sound just doesn't occur in English.

Returning to the rubber-hand illusion: we might have thought that when it comes to feeling being touched, or feeling touched on your right hand, there's no need for disambiguation—that is, no need for contextual clues about where and how you are being touched. After all, don't feelings exhibit their intrinsic quality in their very occurrence? But in fact there is need for interpretation or comprehension. You may be touched on your hand, but you will feel yourself touched on the place where you seem to see yourself being touched. Now, in one sense this is clearly an illusion. After all, you really are being touched on your right hand under the table. But in another sense there's no illusion—or rather, the mechanisms at work in this illusion, if we want to call it that, are those of normal, successful perception. Granted, the rubber hand isn't a part of you. But what explains this is not the mere fact that it is a rubber hand, or that it is not connected to your body. The more fundamental fact is that you and the rubber hand have distinct fates. There is nothing more than a superficial and accidental coordination of your experience with the rubber hand. Your own hand, by contrast, is reliably implicated in your sensory and motor interactions with the world around you and with your other sensory experiences. If it were possible to incorporate the rubber hand into a dynamic of active engagement with the world and the body, then, to that degree, the rubber hand would become a part of you.

We have a very special relationship with our own bodies. Evolution itself is to thank for this, no doubt. But the rubber-hand demonstration causes us to rethink what the special relationship is. It does not consist merely of the fact that my hands and arms, say, channel nerve tissue into me, or rather, into my brain. Connectedness, attachment, contiguity—these are important, but mere connectedness or attachment yields only a superficial explanation of what the body is. What makes connection and contiguity important is that they themselves track coordination and common fate. This is the hand—my hand—whose movements I see when I look. Part of what makes it my hand is that I see it grasping the cup. Part of what makes it my hand is the fact that it is the one with which I grasp the cup. Indeed, there is no specific feeling or characteristic sensation that is or would be the feeling that this is my hand. I feel with it (e.g., the cup is too hot!) and in it (I am being tapped and stroked!). Its "mine"-ness consists in the way it is actively, dynamically, visually involved in my living.

Maurice Merleau-Ponty, the French philosopher, has made just these points. Our lives take place in a setting. This setting—the floor, the walls, the noise, the outside—is the background for whatever activity we are engaged with, such as driving, or walking, or baking a cake. The body, for Merleau-Ponty, shows up as mine (or yours) in just this way as the background condition of my carrying on as I do. These hands belong to me, for it is with them that I break the eggs and mix the batter. As Merleau-Ponty puts it: "The body is the vehicle of being in the world, and having a body is, for a living creature, to be intervolved in a definite environment, to identify oneself with certain projects and be continually committed to them."

Phantom Hands

The rubber-hand phenomenon shows that I can have sensation in an object that is not, in fact, attached to me. I have suggested this is because attachment or connectedness is not necessary for something to be a part of me. Further support for this idea is found in the phenomenon of the phantom limb, a topic discussed in the last chapter. We now know that when a hand is lost to amputation or accident, one does not automatically lose feeling or sensation in what is now an absent and thus phantom hand. This makes perfect sense if, as I am arguing, what makes a hand yours is its involvement with your habits and projects. I still imagine that I hear the rattle of my dog's collar in the night, even though he has been dead more than a year, and I still automatically grope for the light switch on the wall when I enter the room even though I know perfectly well that the switch has been moved. Just as moving the light switch doesn't stop you from reaching, so the loss of a hand does not all at once obliterate the behavioral setting on which your having or believing yourself to have a hand depends. Just as you don't fully feel the absence of your deceased loved one until the time comes when you go to the phone to call him, as Merleau-Ponty once noticed, so the absence of your hand is not real until it fails to be at your disposal when you prepare to reach with it or stop your fall with it. A limb is quasi-present as a phantom limb when the behavioral, environment-involving attitudes and engagements outlive the loss of the limb. Only when you fully adapt to your new circumstances—only when you break the habit of acting with and on your hand—will your ghost hand finally be put to rest.

The psychologist Vilayanur S. Ramachandran has nicely demonstrated this relationship between the sense of ownership of a part of your body and habits and expectations. Patients with phantom limbs frequently suffer; for example, they complain of cramps in the phantom that they can get no relief from because they can't, as you could with an actual hand, work the cramp out by moving. Ramachandran rigged up a mirror box so that a patient's intact right arm would look, to the patient, like the phantom left arm. Now, in the mirror box, it is possible to turn and move and find soothing relief in the phantom. To paraphrase Botvinick and Cohen, in this therapy the phantom hand feels the relief that the eyes see.

Consciousness does not occur in our brains, and the body is not an elaborate vat for an otherwise autonomous brain. To paraphrase Merleau-Ponty, our body is ours—the place where we feel and the means by which we act—insofar as the current of activity that flows toward the world passes through it.

The Body Schema

Psychologists use the term "body schema" to refer to the implicit, practical body plan that enables us to deploy our bodies effectively in movement and action. You don't need to locate your hands before you put them to use in reaching for something, and as a general rule you don't need to pay attention to your body parts (hands, fingers, whatever) in order to use them effectively. In fact, unless you are very much a novice—an absolute beginner learning a musical instrument, for example—you will disrupt your performance if you focus not on the task at hand, or the goal, but on the bodily mechanics of execution. It isn't that the body is not present to us, or that the body is utterly transparent in our dealings with things around us. We do have a sense of our bodies as present. But in the course of engaged activity, the body doesn't make itself felt as an object of contemplation or awareness. Consider: getting through a doorway never presents itself to you as a problem in the way that, say, getting a couch through the doorway does—and not just because doors are designed to be easy for us to pass through. The point is that we don't need to think about our bodies or pay attention to them to act: We do need to think about the couch to carry it from here to there. The body is present in our normal, active, engaged experience in a different way, comparable to the way in which the periphery of the visual field is present as part of the background against which you attend or focus on this or that. We can be more precise still: the body is present schematically as a range of possibilities of movement or action. That's what the body schema is. For example, my arms can be present to me now, even though I am not now thinking of them; the feeling of their presence comes down to such things as my sense that the coffee cup on the table is within reach. To have a normal, well-functioning body schema, then, is for one to have habits of bodily activity; it is for one to have a body ready in the background to serve one's engaged activities. This unarticulated and perhaps inarticulable knowledge of the body's readiness and availability with its natural degrees of freedom of movement supplies the foundation of everything we do.

The body schema can be contrasted with the body image. The body image is a kind of mental picture that we have of ourselves. The anorexic teenager who looks at her emaciated frame in the mirror and feels fat has, in this sense of the term, a damaged body image. Her body schema is likely to be just fine. Her hands and limbs and eyes and head get mobilized in action in the normal way; they are present to her in the background in the normal way. Her problem is that she feels bad about her body, about how it looks and about her ability to control it.

The rubber-hand experiment and the phenomenon of the phantom limb demonstrate that our body schemas can be shaped or altered. Merely losing a limb in an accident doesn't alter the body schema; that's why the phantom lives on. And merely being a detached bit of plastic or rubber is not enough to prevent something from getting incorporated into the body schema.

Extending the Body

One of the simplest ways we extend beyond the limits of our brains and bodies is by using tools. Take the case of a blind person who uses a cane to perceive the ground before him. He feels the texture of the ground at the end of the cane. He feels in the cane even though there are no nerve endings in the cane, even though the cane is a bit of metal or wood. What this shows is not that cane-aided perception is independent of the brain and nervous system; hardly. The point, rather, is that cane-aided perception is not a matter of having feelings either in the end of the cane or in one's hand. The brain and nervous system, insofar as they enable perceptual awareness of the environment, are not in the business of generating feeling; rather, they are in the business of enabling us to interact dynamically with the environment. Our experience and capacities depend on the full character of that skillful interaction. Where we are depends, in significant part, on what we're doing. And what we are—Is the cane a part of me or not? What are the limits of my body? Of myself?—depends on more than the brain alone. Skillful experience with a cane can actually extend the body beyond its strictly biological limits.

These examples show that tool use can modify our body schema. By integrating the tool into a practical repertoire, we are able to remap our expectations of what we can do and so, in effect, we remap the body schema. Drivers can come to feel where the back of the car is as they back into a parking spot, and they can come to sense their contact with the texture of the road through their wheels. In this same way, the baseball glove or lacrosse stick extends the athlete's reach.

As our body schema changes, our relation to the world around us changes, and so how we perceive the environment changes. How big a parking spot looks will be affected by the size of the vehicle you're driving; how steep a hill looks has been shown to vary, depending on the weight of the pack you are carrying. Indeed, it has been shown that the apparent size of the baseball varies in direct correspondence to the hitter's batting average. The better you are hitting, the bigger the speeding balls you are trying to hit will seem! When you're slumping, the balls actually appear to shrink!

These changes in our body schema correspond to processes of neural adaptation. It has been shown, famously, that monkeys using a rake, for example, exhibit enlarged cortical representations of the hand and arm. That is, cells that are sensitive to both the look and the feel of the hand and arm come to treat the rake extension of the arm as if it were part of the body—as if it were the arm.

When our body schema is in this way transformed, our sense

of what is near us also changes. We can map space independently of ourselves, but we can also map it in relation to ourselves. Some regions of space are near us or within reach. Psychologists call this peripersonal space. Other regions are beyond reach. This is extrapersonal space. Transformations of the body schema can bring about extensions of peripersonal into what was merely extrapersonal space. Remarkably, human neuropsychology patients who 'exhibit symptoms of disordered body schema in nearby peripersonal space, but not in more distant extrapersonal space, will begin to show symptoms in extrapersonal space if their dealings with that space depend on skillful tool use. The acquisition of tool-using skills has the effect of increasing the extent of peripersonal space. What was far becomes near.

What are the limits of the plasticity of the body schema? Some degree of plasticity is obviously necessary. After all, through childhood and indeed to a lesser degree throughout our lives, we grow, our bodies change, our degrees of freedom alter. Imagine if one were unable to accommodate the shifting sands of one's bodily geometry and changing abilities. Such a deficit would be devastating. One's body would be like an unfamiliar apparatus: learning to be in one's own skin would be like learning to ride a bike or learning to dance the salsa.

Where do we stop, and where does the rest of the world begin? What these reflections on the body schema show is that there's no principled reason even to think that our bodies stop where we think they do. Parts of me—tools—can be spatially discontinuous with me: What makes them me, what makes them part of my body, is the way my actions take them up. And insofar as I act in and feel with my extended body, my mind is extended too.

None of this is meant to challenge the pivotal role of the brain and nervous system in the whole story. But we can acknowledge this fact without conceding that the brain and nervous system *are* the whole story. For it is only the brain and nervous system in action that allows for a body schema, and we can understand the coming to be of this distributed self only by focusing our attention on the animal or person in action, acting on and acted on by the environment.

Knowing One's Way About

Do you know what time it is? If you're like most people, you'll say yes and then you'll look at your watch. You believe that you know the time now because you know that you can look at your watch. As the philosopher Andy Clark has convincingly argued, we find it natural to think that we know the time even when, in fact, we don't have as it were the actual time in our heads. We know the time when we have quick, easy, reliable access to that information. The actual repository of information about the time—the watch—is on your wrist, not in your head. In this example, the watch functions as a bit of exterior machinery for supporting your cognitive accomplishment of knowing the time.

Most of us can get around in big cities with tolerable ease and without undue exertion. We know how to get from here to there not because we have actually memorized the map of the whole city; most of us, even in cities where we have lived our whole lives, rely on all manner of external markers, signs, landmarks, and maps to find our ways about. My hometown is New York City. If I come up out of the subway in an unfamiliar part of town, I can usually figure out how I stand in relation to things around me. The sign says "Second Avenue." I know that traffic on Second Avenue runs downtown, so I know that traffic is heading south. It's easy enough, then, to turn about and head off to the West Side. Navigation requires that we use our senses, that we pay attention to landmarks, and that we draw on background knowledge too (e.g., that Second Avenue runs downtown).

Now, my ability to find my way around New York City is a cognitive ability, an intellectual achievement, albeit a humble one. But it is one that I possess only given my situation—that is, given the larger context of my access to environmental markers and cues. The environment itself is what enables me to find my way around in it. My understanding, my knowledge, is not something autonomous, something detachable. Rather, it is a skillful familiarity with and integration into the world. Just as we may count on our fingers and calculate with pen and paper, so we navigate with the world. Like my brain, my body, my eyes, and the city itself belong to what enables me to find my way about.

I agree with the philosophers Andy Clark and David Chalmers that there is no principled reason not to think of the wristwatch, the landmarks, the pen and paper, the linguistic community, as belonging to my mind. The causal processes that enable us to talk and think and find our ways around are not confined to what is going on in our skulls. But that is just a way of saying that the machinery of the mind itself is not confined to the skull. The head is not a magical membrane. We are involved with the world around us. We are in it and of it.

Real Presence

We ourselves are distributed, dynamically spread-out, world-involving beings. We are not world representers. We have no need for that idea. To put the point in a provocative way, we are, in Merleau-Ponty's memorable phrase, "empty heads turned toward the world." And as a result of this, our worlds are not confined to what is inside us, memorized, represented. Much more is present to us than is immediately present. We live in extended worlds where much is present virtually, thanks to our skills and to technology.

Consider a simple example but one of enormous richness. You experience the house as having a back, even though you can't see the back from where you are. You do not merely think or judge that there is a back. Nor do you merely infer that there is a back. The back feels present. In my view the back of the house is present to my visual experience of the house, even though it is hidden, because its being hidden does not prevent me from having bodily access to it. Movement of my body in relation to the house—walking around it—will bring it into view; my experience right now of the house has that contour. It is structured by the fact that now, in front of the house, I am related to the back of the house by possible movements. The back of the house is present as absent—that is to say, as out of view.

We can think of this as a kind of virtual presence, but only if we recognize that all presence is virtual in this way—not in the sense of being false, or illusory, or less than genuine, but in the sense that the world is present as reachable, rather than as depicted. And in that sense our relation to the back of the house is not different from our relation to the house's front. We bear a relation to the house: the house itself, the back no less than the front (or the front, no less than the back), is within reach.

Technology increases the scope of our access, and so it increases the extent of what is or can be present for us. My mother thousands of miles away is present, for she is just a phone call away. The funds in my U.S. bank account are available to me here in Germany in this age of electronic banking, and so they are, in that sense, present—that is to say, they feel present to me.

The use of instant messaging provides a striking example of this kind of extended presence. Studies have shown that the use of messaging among teenagers in Japan has transformed the dynamics of social relations. Kids text back and forth throughout the day. They rarely send informative or detailed messages; the informational content of their sendings tends to be minimal. In effect, they are "pinging" each other: letting each other know that they are online, or in reach, or "there." Before school, during school, on the bus home after school: ping ping ping—the reassuring indication that your friend is there for you. In this way, the practice of texting, or instant messaging, creates a new modality of social presence. Just as tools can warp the body schema and connect me to bits of stuff by making that stuff in effect part of me, so the pinging action of instant messaging can enable individuals separated by space across Tokyo to operate socially within each other's sphere. They are virtually present to each other.

A case can be made that joint presence in an actual shared physical space is the best kind of presence. We are embodied biological creatures and evolution has conditioned us perfectly to fit into actual physical niches. We are naturally attuned to the physical environment and to each other. Physical contact is polymodal: we hear the words and see the facial expressions and feel the heat of each other's breath and jointly attend to what is going on around us. Virtual worlds, in comparison, are thin in a distinctively digital way. I don't mean that virtual presence is merely illusory. It is real presence extended by new and different methods. There is something vaguely disturbing about the thought that passively sitting at a computer, reading and typing, could be a genuinely active, outgoing, socially engaged mode of being. But that is exactly what it is.

The Case of French Air Traffic Controllers

Air traffic controllers in France and around the world use paper strips to stand for airplanes. As a plane enters a controller's airspace, the clatter from the printing of the paper strip alerts the responsible controller. He manually annotates the strip and places it in a tray or on a special table with other strips standing for planes in the sky for which he is responsible. As the courses of the aircraft are tracked by the controller, the paper strips are annotated, handled, their positions shifted on the tray. When an airplane passes out of one controller's airspace and enters the space of one of his colleagues, he hands (or throws) the strip to him. This handing off of the airplane proxy symbolizes, for the controllers, the handing off of responsibility; the paper strip in turn carries all the information about this plane that the next controller needs—supplemented by radar information, of course—so that he can see the plane on its way safely through the airspace. There are small differences in the way individuals note the paper strips and in the way coworkers in a given air traffic control station handle the strips and, more generally, process the airplanes; there are larger differences from country to country. But most air traffic controllers everywhere in the world carry out a work practice something like what I have described.

Wendy Mackay and her colleagues in France made a careful study of the role of paper flight strips in air traffic control. Part of what interested them was the fact that air traffic controllers have actively resisted any efforts of engineers and policy makers to upgrade the technology of air traffic control. Efforts to digitize the tracking of airplanes were rejected by the air traffic controllers. They liked their low-tech paper solution. More important, at least when Mackay was carrying out her research in the nineties, not one single fatality in France had ever been the result of error on the part of air traffic controllers. Indeed, it was fear that changes in their work habits would make plane travel less safe rather than safer that led the controllers to be resistant to innovation. Because air traffic controllers are the ones whose careers are on the line, it is reasonable that due consideration is given to what they want.

What Mackay and her collaborators observed was that the paper-strip system, although low-tech, was in fact an extremely well-adapted technology. The paper strips were not merely files of information about the airplanes: they were airplane proxies. Keeping track of the planes was made easier by keeping track of pieces of paper in very much the way that figuring what words you can make out of your Scrabble pieces is made easier by shuffling around and trying out possibilities with the pieces. The paper strips did indeed record information about the planes, but the information on each paper strip was annotated by the very controller with responsibility for the plane. The act of writing on the strip—the fact that the strip was handled as well as seen and read and set down on a tray in relation to other strips so as iconically to represent the airplanes—turned out to facilitate greatly the basic job that the controllers faced: keeping track, in real time, of many different aircraft coming and going, ascending and descending, carrying the precious cargo of human passengers. Moreover, Mackay learned from her observation of the controllers at work, because the paper strips were stored and manipulated in full view of the other controllers, the strips became a collectively shared analog device for letting the whole crew of controllers keep in touch with what was going on in the regions of airspace surrounding those planes for which they were directly responsible. Conversation among the flight controllers as well as radio communication with the planes themselves, the sound of the printing of new strips as well as the placement, handling, and handing off of the strips, formed a shared environment in which the controllers, as a group, were able to accomplish their task effectively.

Two crucial observations can be made about the system of paper strips. First, the paper strips played a critical and indispensable role in the cognitive repertoire of the controllers. They couldn't succeed in their task without them. Their active engagement with the paper strips was a kind of cognitive involvement with the planes. They thought about the planes with the strips. Second, the mission of the controllers was achieved collectively: the noisy, cluttered workspace, with shared tables for paper strips and shared radars, provided for what has been called a kind of socially distributed cognition.

According to the modern conception, we are thinkers, individualistically achieving an internal representation of and understanding of the world—or at least of *a* world that is, or seems to be, beyond our personal limits out there around us. Mind arises out of the workings of our individual brains; brains are our mind organs. But not all experience and cognition is in this way detached, contemplative, individualistic. Much of our cognitive lives—vast stretches of our linguistic lives, our cooperative work lives, as with the air traffic controllers—requires not only landmarks and tools (such as language or paper strips) but also other people.

Extending the Mind

We have been considering ways in which body and, in effect, mind are extended by the use of tools and other artifacts. New technologies—"new media"—provide important and challenging examples. But there is no more robust technology for the extension of human thought beyond the limits of the internal, private resources of an individual than language.

Simple examples are readily at hand. Without language it would be impossible, surely, to think about the question of whether I had breakfast the morning of November 1, 1974, for it is language itself, with its implicit system for dating, counting, and conceptualizing time, that provides the very medium in which such a thought can take form. No nonlinguistic brute could fashion that thought—that is, that particular relation to the world—for it is a relation that is linguistically structured. Or consider that although people had nothing long before the Arabs invented a sign for zero, it is only with the advent of this new symbol that the integers could be born. There can be no serious doubt that the Arabic system of numerical notation is vastly superior to the Roman system; this fact had great implications for what ancient peoples could accomplish in mathematics. Even today there are very different linguistic-conceptual methods for representing numbers. The French call ninety *quatre-vingt dix* ("four-twenties ten"). It would not be surprising to learn that the way of conceptualizing what English speakers would instead think of as "nine tens" makes a difference in our respective arithmetical competencies, at least in some situations. According to the French neuroscientist Stanislas Dehaene, all European languages are at a disadvantage when compared with Mandarin; he argues that the Chinese are on the whole better and faster at mental arithmetic and that the explanation lies in their linguistic system for talking about numbers.

The point is that just as a rake extends our reach and has the potential to extend our body schema, so language extends our capacities for thought and therefore can be treated as extending our mind schema. Insofar as language is itself socially manufactured and shared by linguistic communities, then to that extent our cognitive powers require for their very exercise the existence of a sociolinguistic environment. Our minds cross out of the skull and get supported in a shared sociolinguistic scaffolding.

We need to appreciate the ways in which our mental powers—for example, our abilities to think about numbers and perform calculations and engage in abstract reasoning—are not, as it were, achievements we reach in isolation, each of us within his own brain relying on his own inner fortitude. We use tools to think, and sometimes those tools are not only external (that is, literally outside the head) but frequently shared and collective (like language and the paper strips of the air traffic controllers). The machinery that enables us to think and calculate and do mathematics may be partially inside us. But it is not all inside us.

Meaning Isn't in the Head

Skeptics may object that even if it is true that language is a collective cultural instrument, each of us internalizes language. That's what it is to know a language. Not every person knows every language. When you learn a language, you learn the system of rules that allows you to think and represent and reason within that language. Thinking, reasoning, etc., all takes place inside of us. The phenomenon of language, then, correctly understood, doesn't provide any evidence in support of the idea that our minds do not reside in our heads.

This objection relies on what deserves to be called the classical conception of words, meanings, and language. According to the classical conception, we use language to describe the world, to make true statements. Words refer to things or qualities. To know the meaning of a word is to know what it refers to. "Water," for example, refers to water or to H_2O—that is, to the transparent liquid that flows through our rivers and streams and runs out of our taps. "Gold" refers to a malleable yellow precious metal. "Beech," "elm," and "oak" are names of trees. To know a language is to know the meaning of its words—to have those meanings in mind. Those meanings fix what you are talking about when you use language to talk.

In the last sixty years this classical picture of language has been demolished in philosophy. I couldn't tell an elm from a beech to save my life, but that doesn't stop me from making true statements about elms, such as that elm trees in America are being killed off by Dutch elm disease. How can I succeed in using the word "elm" to refer precisely to elm trees even though I my-

self am unable to pick out the elms? Answer: I am not myself, in-
dividually, responsible for making my words meaningful. They
have their meaning thanks to the existence of a social practice in
which I am allowed to participate. It is crucial, for the whole
practice to keep going, that there be experts who can tell elms
and beeches apart. But, thanks to the social division of linguistic
labor—as Hilary Putnam, who is responsible for these ideas,
calls it—there is no requirement that each individual carry the
burden of securing meaningful reference. We rely on others.
And we can do that because meaning is not something internal:
it is not internal to me; it is not internal to the experts. Meaning
depends on the practice, in very much the way that the powers
of the rook in chess depend on the practice.

None of this should be taken to mean that you don't need to
know what you are talking about in order to talk! Understanding
is vital. But Putnam's ideas about language—and he was building on Wittgenstein's ideas—give us a way of appreciating that
even understanding (e.g., knowing the meaning of words like
"elm" and "beech") is not a matter of having in one's head a rule
that fixes how to use the word. Rather, it's a matter of being able
to use the word correctly (that is, of being able to participate in
the practice)—being able to do that counts as knowing, for all
practical purposes, what an elm tree is.

Beyond the social division of linguistic labor, Putnam also believes that linguistic meaning depends on one's being embedded
in an environment. Even before we learned that water was
H_2O—that is, even before there were experts who could differentiate water from superficially similar liquids—we were able to
use the word "water" to refer to water because, after all, it is *this
stuff*—this very stuff we drink and bathe in and cook with—that
we refer to when we talk about water. It is our everyday inter-
course with water that makes it the very stuff we mean when we
use the word. The word is stationed in a practice and the practice is one that makes use of actual, bona fide water.

The point can be extended further. Granted, to learn a language is to acquire knowledge. But to a very great extent linguistic knowledge isn't knowledge that such and such is the case; rather, it is the acquisition of a set of abilities to use words and handle things. In a range of cases, linguistic knowledge depends on our active engagement with our surroundings. In no sense, then, does the fact of language provide support to the defenders of a more Cartesian, individualistic linguistics.

The Secret Lives of Snails

One of our cherished conceptions of ourselves is, well, that we have selves, kernels of "me"-ness that dwell inside and control and guide and govern our movements and actions. This idea has come steadily under attack in twentieth-century thought. Freud, for example, argued that the conscious mind does not govern anything; at most it makes up stories about itself. And this is a familiar idea in contemporary cognitive science. Mainstream neuroscientists, it seems, haven't heard the news and tend to incorporate a simple conception of the self into their basic assumptions about the brain and its place in our lives. So, for example, a guiding metaphor is that the brain is Mission Control; it is Headquarters. Brains monitor, gather, and assess "intel," send and receive transmissions, commands, warnings; brains manage that tightrope between sensory input and behavioral output that is animal life.

The Mission Control metaphor is distorting. To begin to see what's wrong with it, consider remarkable findings about the neurocognitive lives of common marine snails (*Aplysia californica*). If you touch a marine snail, it will withdraw from the touch. It pulls back the way you might pull back from a hot object. The snail will relax in the face of repeated touching; it gets used to being touched and no longer automatically withdraws.

This habituation of the snail to touch, a kind of learning and memory, can be contrasted with another sort of learning called sensitization. If a snail receives an electric shock instead of a gentle stroke, it snaps back. Repetition of shocks quickly teaches the snail to be wary of any contact; the snail will now pull back sharply from even the most gentle contact. It isn't hard to relate to the snail. Think of when you get a shock from static electricity; despite the fact that the static has been discharged, you'll be tentative in reaching out to take the doorknob.

Eric Kandel and his collaborators, first at New York University and then at Columbia University, investigated the neural foundations of this sort of learning in sea snails. The sea snail, they realized, is an ideal subject for this sort of study. It has only about 20,000 neurons, and it is pretty easy to see them and study what happens as the snail negotiates this sort of situation. Simply put, what Kandel and colleagues learned was this: Sea snails have sensory neurons that respond to touch and they have motor neurons that produce movement. The sensory neurons and the motor neurons are wired together; stimulation of the snail by touch activates the sensory neurons, which in turn directly activate motor neurons, producing the withdrawal response. What happens in habituation is this: Repeated harmless touching causes the strength of the connection between the sensory cells and the motor cells to weaken. Repeated stimulation changes the physiology. If repetitions occur often enough, the change in the physiology can be as good as permanent. (Think of the way you don't feel your clothes.) Sensitization is the reverse process. Painful contact produces a strengthening of synaptic connection between sensory input and motor output.

The sea snail learns; it remembers; it modifies its behavior in light of this learning. But now ask yourself: Where is Mission Control in this story? Where does this learning take place? Where is the decision made to snap back from a touch, or to relax in the face of repeated touching? Answer: There is no Mission Control. The learning is itself distributed across the nervous system of the animal. The snail learns; the embodied nervous system of the animal as a complex network subserves this process.

More remarkable still, what drives the whole process of physiological reorganization is the stimulating environment. The neurophysiological adjustments that take place depend, of course, on the animal's genetic makeup and on the nature of the molecular chemistry that governs the processes whereby synaptic connections strengthen and diminish. But the arc of the changes that take place is governed by the character of the environment's action on the animal. The changes that occur are a function of the environmental situation.

Let us ask: What explains the animal's learning and memory? To understand the animal's achievement we need to look to the ongoing dynamic between the animal and its nervous system on the one hand and the environment on the other.

Where does the sea snail stop and the rest of the world begin? At first blush, the obvious answer may seem to be that the relevant boundary is the surface of the snail's body. But a little further consideration reveals that the sea snail is what it is thanks to the way it is bound to, affected by, and coupled with a specific situation. The world acts on the snail; the snail responds; how it acts is shaped by how it was acted on; the snail is a vector resulting from distinct forces of the body, the nervous system, the world. Its past history *in the environmental context* and its ongoing dynamic exchanges with the environment make the sea snail what it is.

Where Do *You* Stop?

Are we so different from the sea snail? You may think so. After all, *we* are conscious. Our actions are not simply synaptic re-

sponses to sensory stimulation. We perceive; the world shows up
for us; we reason and plan; our actions are shaped and driven by
what we know as well as by what we feel, need, and want. In
Chapter 5, I'll argue that this is the wrong way to think about
consciousness and its role in our lives; the point then will be that
consciousness is not this kind of reigning deliberative awareness.

For now, though, what is on the table is the Mission Control
conception of the brain. Our nervous systems are much more
complicated than that of the sea snail, to be sure, as are the myr-
iad ways in which the world can act on us and provide opportunities for us to act. But like the snails we are bound to the world,
and what we are and what we can accomplish depend on what is
done to us as much as on what we do. We share this with the
snail: like the snail, we are not autonomous. We are in the world
and of it.

A simple and perhaps familiar fact can help to drive this
home. A striking thing about mammals is that we tend to be born
immature. We achieve our maturity, with all its attendant capacities, only after a long infancy. But what is infancy but the process
whereby the environment acts on us and makes us into what we
become (as discussed in Chapter 3)? Take the mammalian visual
system: If an infant were prevented from using his eyes—for example, if his eyelids were stitched shut, or if he were reared in the
dark, or if he had cataracts—the lack of stimulation would prevent the establishment of the rich arbor of neural interconnec-
tions that is in fact necessary for mature vision. It is experience
that is required if the infant is to acquire the neurophysiological
machinery to have mature experience.

This kind of developmental plasticity may be compatible
with the Mission Control picture. It could be taken simply to
add a wrinkle to the story: the infant brain is plastic; interaction
with the environment is necessary if it is to become the kind of
Headquarters that it can become. But it is much more com-

pelling to think that what all this reveals is that the brain is no more in charge of what you do than a surfer is in charge of the wave he's riding. Brain, body, and world form a process of dynamic interaction. That is where we find ourselves.

Conclusion: We Are Involved with the World

Landmarks, tools, shared places and practices, belong to the machinery of our being. We are partly constituted by a flow of activity with the world around us. We are partly constituted by the world around us. Which is just to say that, in an important sense, we are not separate from the world, we are of it, part of it. Susan Hurley said that persons are dynamic singularities. We are places where something is happening. We are wide.

Many of us were taught to believe that habits are bad. They are there to be broken. In the next chapter I try to convince you that habits are in fact essential to our mental lives. Because habits are highly dependent on the environment, they provide a striking illustration of the way in which our manner of being implicates our ongoing dependence on the world around us.

HABITS

Wanderer, the road is your
footsteps, nothing else;
wanderer, there is no path,
you lay down a path in walking.
In walking you lay down a path . . .

—Antonio Machado
(translation by Francisco Varela)

Scientists have tended to think that to have a mind like ours, one must be able to think and calculate and deliberate as we do. In fact, to have a mind like ours, what is needed are habits like ours. Habits and skills are environmental in the sense that they are triggered by environmental conditions and they vanish in the absence of the appropriate environmental setting. We can reject the idea that we are autonomous islands of decision making, acting on the basis of careful scanning and sound judgment. Our nature is more intimately entangled with our environment than that.

Creatures of Habit

Human beings are creatures of habit. Habits are central to human nature. Roboticists should take heed; they've directed their energies to making clever robots—robots that can make chess moves or avoid obstacles. A better goal would be to make robots with habits. My hypothesis: Only a being with habits could have a mind like ours.

Traditional approaches to the mind in cognitive science have failed to appreciate the importance of habit, for they start from the assumption that the really interesting thing about us human beings is that we are very smart. We are deliberators, we are propositional, we use reason. We perceive, we evaluate, we plan, we act. We are the rational animal. This idea has roots in Plato's view that the rightly ordered person subordinates emotions to reason. A similar picture can be found in Descartes, who insisted that each of us has an intellectual responsibility to call into doubt even our most mundane beliefs ("This is a piece of wax," "Those are people walking by outside the window") and to rebuild our system of knowledge piece by piece from the ground up. Descartes seems to have thought that we could, at least in principle, achieve in this way a sort of perfect rational mastery over our world conception. What these views have in common—and what they have bequeathed to cognitive science—is the idea that we are, in our truest nature, thinkers. It is this intellectualist background that shapes the way cognitive scientists think about human beings.

According to the intellectualist conception, we are habit-free. Our distinctive nature reveals itself precisely in the fact that we perceive, we evaluate, we decide, we plan, we act. We are free precisely because we rise above mere habit and act from principles. Now, there is something to this. There is a value in the freedom that comes from deliberation and decision; it is valuable to rid ourselves of prejudice. But here's the sticking point: Judgment, deliberation, decision making, always take place in a context, a setting. There is no such thing—Descartes' fantasy notwithstanding—as deliberation and judgment all the way down. Making judgments, applying categories, interpreting—all of this requires that one's terms and concepts are settled, at least by and large. I may decide that the building before me is Georgian based on the fanlights, but in making a careful judgment of that sort I take for granted what a house is, or what a window is. Likewise, if I am working on a mathematical problem, I may be pushing my understanding to its limits, but this is only possible because of my confident mastery of the more basic skills (such as counting) on which I depend. The fantasy of rational emancipation is just that—a fantasy—at least if it is meant to suggest that there can be a presuppositionless form of mental life.

The real problem with the intellectualist picture is that it takes rational deliberation to be the most basic kind of cognitive operation when in fact thinking and deliberating are themselves the exercise of more basic capacities for skillful expertise. And the hallmark of expertise is its fluency: it is engaged and, precisely, nondeliberative; the expert eschews just the kind of distanced, careful contemplation that, according to the intellectualist, is definitive of our truest nature. The intellectualist portrays human beings in the course of their lives as inevitably novicelike and always, for that reason, as unskilled newcomers who are in effect alienated from the world around them. For Intellectual Man the world shows up as strange and objectified, something to be figured out, interpreted, analyzed; we are forever in the role of Mr. Spock, asking ourselves, "What is this strange Earth custom?"

From Novice to Expert

Let's look more closely at expertise. When you learn a new skill—for example, hitting in baseball—you have to pay careful attention to the mechanics of what you are doing. You learn to hold the bat and swing just so, you focus on keeping your motion level and rotating your wrists in just the right way. Psychologists have demonstrated experimentally that as a rule, novice performance improves with this kind of focused attention on the mechanics of the task. You improve your performance when you pay attention to yourself and what you are doing.

Things are just the reverse when it comes to the expert. The expert's performance, it has been shown, deteriorates if he focuses on the mechanics of the task. Expertise not only allows for but in fact requires that the expert turn his attention elsewhere—for example, to tactics, or to figuring out what sort of pitch is likely to be thrown next. The expert's performance flows from an engagement with the larger activity that is necessarily unavailable to the novice. Paying too much attention to what he is doing, to the mechanics of the task—in other words, behaving too much like a novice—will interrupt the flow and likely cause the expert to choke.

That novices and experts have qualitatively different manners of involvement with what they are doing has also been confirmed by neuroscience. It has been shown, for example, that highly trained experts—musicians, athletes, etc.—show a decrease in the overall level of brain activation when they are engaged in the performance of their skills compared to beginners. In a way, it is almost as if the better the player, the less there is for the brain to do! For the experienced player, the task takes over. Likewise, experts display a general reduction in muscle activation. The physical systems drawn into play by the expert show economy and elegance. To cite the title of an article on the neuroscience of expertise: "The mind of expert motor performance is cool and focused."

In the Beginning Was the Situation

It is abundantly clear that we as individuals do not stand as novice and stranger in relation to the world around us. There is something deeply misguided about this intellectualist conception, and therefore we make a big mistake if we suppose that the closest analogy of our relation to the world is that of the computer or robot; the latter are designed, precisely, to embody the kind of detached, intellectual attitude—building up internal pictures of the scene, making plans, and executing them—that the intellectualist wrongly thinks is the defining feature of our mental lives.

What is no less clear is that the intellectualist conception—according to which deliberate judgment is the very paradigm of our minds at work—is wrongheaded even as an account of our intellectual activities. Here, too, in intellectual domains like chess, mathematics, talking, and reading, we find the contrast between the expert and the novice; here, too, we find that expertise requires precisely the absence of care and deliberation that the intellectualist wrongly takes to be the hallmark of our mental lives.

As an example, take the experience of learning a second language. In the beginning you learn words, one by one; you struggle painstakingly to pronounce them correctly; you memorize the rules for their declension and conjugation and for combining them with other words. To begin to speak in a language that is not your own, you must—in the beginning at least—spend a lot of time paying attention to language itself; that is to say, you have to be distracted from what is normally of interest to us when we speak—namely, what we are talking about and who we are talking with! It is often said that it is much harder for adults to learn languages than young people. This is no doubt true. I suspect that part of what explains this is the fact that younger people are simply more willing to suspend their engagement (with tasks, people, work) than older people; or perhaps younger people are just less engaged. A nineteen-year-old will be thrilled to spend the day in a café practicing his chitchatting skills. An adult is more likely to need or even want to be at work!

To the degree that the novice must focus on language itself

and its rules, the fluent speaker is unable to do that. She has probably forgotten the rules, tables, and tricks that were so important when she was a beginner. Moreover, nothing will cause one's fluency to falter more than focusing on the words and sounds instead of on what one is saying. (I would expect that this is true even for actors.) To know a language is to be able to leave any concern for the rules behind, thanks to the fact that the rules themselves have been mastered. Crucially, the expert speaker/ hearer of a language differs in her relation to its forms and possibilities qualitatively from that of even the very best learners; knowing a language is never a matter of applying the rules that you learned as a beginner, only more rapidly and with fewer errors.

In general, mastery in a language does not confer a special ability to think or talk about language or to explain its regularities and principles, just as being a good guitarist doesn't carry with it the understanding of how best to teach someone else to play. In many societies—our own included—this unquestionable fact can be concealed from view. For our very picture of the place of language in our lives is shaped not only by our daily linguistic engagement but also, and thoroughly, by the ideology or theory of language that we learn in school. We are taught about language, and so we find it natural to think of languages as all-purpose, context-independent symbolic systems for naming, describing, truth telling, and informing. We are taught to use dictionaries, and so we find it natural to think that for every word there is a distinct meaning, or perhaps a list of distinct meanings. When we learn a second language, we are taught to translate sentences from our native tongue into that language and vice versa. And so we learn to think of languages as standing in determinate relations to one another. It can be difficult or impossible to realize that, for someone who has not been indoctrinated into such a conception of languages as intertranslatable and documentable, the very idea of translating from one to another can seem as strange as "translating" football into baseball. In most of the world, people live in densely multilinguistic environments. Indeed, the very idea of one people/one language is a cultural invention of the nineteenth century. But even though a person in Zinder, Niger, is likely to speak several languages—at home she speaks Fulani; in the market she speaks Hausa; she listens to the news in French—the question of translating between these languages does not arise. Why would you speak Fulani in the market? And what possible reason would there be to speak French in the home? Languages are not abstract symbol systems, or at least this is not all that they are. Languages are aspects of engaged human living.

This fact—that language and activity are interwoven; that languages are particular aspects of our engagement with situations—is something we constantly encounter. Teenagers use language differently from their parents, and it would be a mark of a teenager's linguistic and social sensitivity that he would modify his speech when speaking to an adult, and perhaps without even noticing that he does so. (Why would one even think of using teenager language with an adult?) We modulate our language throughout the day, depending on whether we are buying bread in the bakery, teaching a class, or speaking to a police officer or a foreigner. Indeed, our language modulates in response to our interests in and expectations of the person with whom we are speaking. How I use words to refer to distant horizons of significance will depend on what I take my interlocutor to know. Baseball fans, computer technologists, and financiers employ jargons that belong to their domains of interest and expertise. To learn about computers or about cricket is, in this way, to learn a new language.

To know a language, then, as a general rule, is not to have knowledge about language, although it will ordinarily be accompanied by some such knowledge—just as knowing how to play baseball isn't a matter of knowing about baseball. Mastery of a language is a cognitive achievement; it exemplifies intelligence and the use of our minds. But it is not a matter of having memorized this or that or of having internalized a complex symbolic system. A language user is, to the extent that she is expert, the participant in a specific social practice; crucially, she is the participant in a social practice of which language forms only an aspect.

The Language of Chess

Scientists in the intellectualist tradition of Plato and Descartes treat thought as a kind of computation. And they think of computation as a matter of deliberate, painstaking calculation. Insofar as we are thinkers, we are all, as far as this tradition is concerned, biologically evolved computers.

Consider, for example, the standard approach to chess in cognitive science. Chess is a complicated and challenging game; many people play chess, and some play it very well. From the standpoint of the intellectualist conception of the mind, this is an impressive fact, for chess presents a daunting computational challenge. The chess player must select, from among an astronomically large number of possible legal moves, the single move that most optimally serves to realize the goal of victory. To do this, the player must, in effect, form an accurate representation of the state of play and then work out or calculate the consequences of possible moves; he must then evaluate those consequences in light of their overall desirability; and he must do this under time pressure. Moreover, the problem arises in a more or less new form every move! To play chess, or at least to play it well, one would have to be a computer! The fact that we play chess so well shows that we *are* computers. We may not know it, but our minds are computing away within us, subconsciously, telling us what moves to make. The fact that computer programming has been so successful in the domain of chess is taken to be further confirmation that we and the computer are on a par; we face the same problem in the same way when we play chess.

Next stop: a full-blown research program for experimental psychology. Computer programs are algorithms. If we are computers, then it must be possible to uncover what sort of program we are running. What sort of symbol system is our chess program playing on? This is something that can be investigated by experimental psychology. Perhaps what explains the recent superiority of artificial computers is simply that their programming is better than ours. Or perhaps it has more to do with the hardware: maybe our processors are just not fast enough, or large enough.

This is the litany of many scientists, for whom what I have just offered is gospel truth. But there is reason to be suspicious of this whole line of argument. We human players of chess don't need to select the good moves from among the nearly infinite possible moves. For anyone who understands chess will know that very few moves are even relevant to the play at a given configuration. On top of that, much of the time the position on the board forces our moves. Even if there are alternative ways of responding to an opponent's move, most of the time there will be, at most, only one or two moves worth considering. The point is that the competent chess player does not face the computational problem of evaluating moves from among the potential infinity of possibilities. Nor does the competent chess player face the computational problem subconsciously or in the brain. The problem does not even arise. Chess is an artifact; it has a point. We play to win; we play against people; we play against people who share our understanding and our goals. The game itself—

pieces, board, rules, customs, history, etc.—provides the environment in which we find ourselves when we are playing chess.

Granted, chess is a mental sport and it tests our understanding and intelligence. If we give up the idea that the plodding,
tentative deliberation of the novice is our paradigm of understanding and intelligence—if instead we appreciate that genuine
competence if not true mastery gets expressed precisely in the
absence not only of the need for but the very possibility of deliberation—then we can appreciate that chess programs take the
chess itself out of the game, and with it the understanding or intelligence that makes chess so challenging. Chess machines are no
more chess players than novice batters in instructional leagues
are full-fledged players. In fact, they are less so.

The Language of Thought

The same points can be made about language. The standard approach to language in contemporary "scientific" linguistics would
have it that our basic competence, when we are competent
speakers of a "natural" language, is in knowing rules for combining words into grammatically well-formed strings; our basic
competence as listeners is in assigning meanings to sentences
spoken by others on the basis of an understanding of the meanings of the individual words they use and the rules governing
their combination. Language use, it is widely supposed, depends
on our ability, or rather, the ability of our brains, to analyze, break
down, and decode strings of sentences quickly and reliably.

This sort of approach to language rests on a systematic ne-
glect of the actual phenomenon of language in very much the
way that the computational approach to chess ignores the actual
character of playing chess. Consider that most of what we say
and hear we've said and heard before. Conversation rarely leads

us out into the untamed wilds; most of the time we are on the school playground or following the paved walkways around the pond or the well-worn dirt paths to the benches. Our linguistic worlds—like the rest of our worlds—run along trails made through repeated walking. And as with walking, it is hard to step off the trail; like water rushing to the lowest point, linguistic thought follows a path to a lowest basin, a course of attraction that we can barely resist. What conversations have you had today? With your spouse, or your kids, or the man who sells newspapers on the corner, or the receptionist out front when you were collecting your mail, or your son's nursery school teacher? The vast majority of what we say and hear is what we say and hear every day of our lives.

This is not a cause for alarm. One of the very many false ideas about language is that its primary function is to express information or communicate thoughts. Speech has many functions, but surely a large part of it is more like the grooming behavior of chimpanzees or the shepherding behavior of dogs than it is like reasoned discourse among parliamentarians. We bark so that our kids get out the door in time to get on their bus and so that they feel safe and loved; we purr so that our colleagues and coworkers know we're on the job and ready to be called on. The bulk of what we say and do each day is more like the grunts and signals baseball players use to indicate who'll catch the pop fly than it is like a genuine conversation.

Linguists tend to be impressed by what Noam Chomsky has called our linguistic creativity—our ability, that is, to understand and produce a potential infinity of sentences we have never heard before. Our knowledge encompasses the infinity of well-formed sentences, sentences of no fixed length formed by combining a finite number of words in accordance with a finite number of rules. This is an even bigger computational problem than that supposedly faced by the chess player, and it's one that even the least articulate among us can solve with mastery. The challenge of linguistics is to figure out how we—or our brains!—manage this task.

But we don't manage this task! There is no need. As we have already observed, much talking is more like barking than it is anything like what the linguists have in mind. Moreover, a good part of what enables me to understand what you say is that I already know what you are going to say before you say it! I never even encounter the problem of needing to assign a meaning to your utterance on the basis of prior knowledge of the words and the rules for their combination. That problem just doesn't arise. For one thing, you and I are (usually) in the same situation. We jointly look at the picture or we both notice the funny sign. Our conversation flows from what we're doing and where we are. We talk about this (thing visible to both of us) or that (thing we both saw yesterday, or are afraid of seeing, or are curious about). What we say refers to and is a reaction to what is happening to us jointly. As an aside, this is what makes the use of cell phones while driving so very dangerous. To carry on a conversation, you need to posit, as it were, a shared contextual setting. A driver, however, has a special requirement to pay close attention to the actual physical environment. The danger arises from dividing attention in this way. It is notable that the conflict doesn't arise (at least not to the same degree) when the driver converses with a passenger. What explains this is the simple fact that now the conversation can unfold against the background of a single shared environment.

It almost never happens that you are simply given a person's words out of context, with no idea in advance of what is intended, what is meant, what speech act is being performed, what ends are being served. I submit that if we were to find ourselves in such a situation, then we would almost certainly be unable to understand even a bit.

I actually had a nice experience of this fact the other day. I was sitting with my six-year-old son on the S-Bahn in Berlin. We'd only just arrived in Germany and my son, as of yet, knew no German. Across from us sat a man reading the newspaper. On the seat next to him, sitting for all the world like a passenger, was the man's dog. That sight alone was amusing to us. My son leaned over and said to the man in English, "Is your dog friendly?" The man looked at August with confusion. "Is your dog friendly?" August repeated. More confusion. I then said, in German, *"Ist er freundlich?"* At those familiar words in the man's native German, his brow cleared and he turned to us and said, in perfect English, "Indeed, he is most friendly!" And he added, in English, "I must be going deaf!" Of course, he wasn't going deaf: he couldn't perceive my son's words because he was unable, in that context and at that early hour of the morning, to find anything remotely intelligible about the sounds my son was making. I don't think the fact that my son's English is that of a six-year-old was particularly a factor. The man knew English but didn't expect English in that situation, and so he was *as good as deaf*.

You may have encountered the same problem speaking on the telephone. My first name is somewhat unusual and I find that it is, without putting too fine a point on it, impossible to convey over the phone. People hear what they expect to hear; since they never expect to hear "Alva," they don't hear it. The solution, in a case such as this, is to spell the name out. But we encounter the same problem again. When there is no shared context and all you want to get across are the bare letters or words, you have no alternative other than to import a context by using standard sound-introducing devices such as "Alpha," "Lima," "Victor," "Alpha"; "November," "Oscar," "Echo". Without doing so, we are confined to the mere speech noises themselves. And these are, to a surprising degree, undecipherable! (In German, it is normal, on the telephone, to say *"zwo"* instead of *"zwei,"* because *"zwei,"* two, sounds too much like *"drei,"* three.)

Neuroscientist and anthropologist Terrence Deacon has made an important observation that bears on the approach to language taken by linguists working in the Chomskyan (Cartesian, intellectualist) tradition. According to that Chomskyan tradition, language—even more so than chess—presents us with a daunting computational problem. The young child must figure out how to use language in the course of just a few years and on the basis of what is widely asserted to be poor linguistic data, owing to the performance errors of everyday speech. Somehow, on this basis, the child figures out the rules that demarcate the domain of the grammatical and the ungrammatical. This is an achievement one would attribute to great scientific minds, yet it is accomplished by every normal child in the world.

Deacon offers a very different suggestion. Language, Deacon points out, is like a graphical user interface such as the Macintosh computer operating system or Windows. There's no mystery about why we find it so easy to use these GUI systems. They were designed by us to be used by us, and crucially they were designed with an eye to what we find easy and manageable. And so with language. Language may be an immensely complicated symbolic system, but we didn't just happen upon it: it doesn't just happen to be that we can figure out (miraculously!) how to use it. We ourselves have built language—collectively, over thousands of years—precisely to be a way of collaborating and communicating that is easy for us.

The Many Faces of Expertise

Intellectualism does a bad job of modeling even very intellectual abilities such as the ability to use a language or play chess. This is because expertise, even in intellectual domains, requires the absence of a certain kind of deliberative distance. Context and past experience enable the expert to simply know his or her way around. This "knowing one's way around" is a cognitive skill, no doubt; indeed, it is an achievement of the highest order. But it is not one that conforms to the intellectualist's deliberative model. The expert isn't someone who simply uses rules quickly or unconsciously; the expert is someone for whom, a good deal of the time, the question of rules does not even arise. For comparison, it may be possible to come up with rules that would allow you to figure out how another person is feeling. One such rule might be: "If the person is sobbing, then she is sad." That's obviously very crude: one reason is that you might think that you would already have to know that a person is sad or distraught in order to identify what she is doing as sobbing. But putting that sort of concern aside, what is clear is that being sensitive to another person, having empathy, is not just a matter of knowing rules like this. Even needing to raise the question "What emotions does this behavior express?" requires that one establish a certain distance from the other that is in fact incompatible with real sensitivity or empathy. There's no room for the rules; there's no place for deliberation. And so, with all the more reason, there is no place for the swift unconscious application of rules.

One area where human beings demonstrate prodigious skill is in their perceptual sensitivity to faces. Even the least clever among us has a high level of skill at recognizing friends, family, and coworkers. In fact, there isn't really any other kind of object or quality to which all humans show such highly cultivated attunement. We can all tell a car if we see one, or a horse, or an X-ray, but very few of us have any skill when it comes to reidentifying a particular horse or picking out the specific breed of the horse standing in front of us. To the uninitiated, at least, all horses basically look the same. Strikingly, when it comes to faces, we are all experts. No special training or years of experience are required, it seems.

One might explain this distinctive human facility with faces by noticing that each of us does have years of experience and special training. We find ourselves among others from the very beginning. As a matter of fact, a sensitivity to faces is a prerequisite for normal human communication. As noted in the last section, part of what enables me to understand what you are referring to is that I can see where you are looking. And it is by paying attention to your face that I learn where you are looking, how you feel, what you want, and so on. So just as it isn't surprising that pig farmers get very good at telling their pigs apart and radiologists can easily pick out the telltale marks in an X-ray, it isn't really surprising that, with regard to faces, each of us is an aficionado.

A very different account of our extraordinary face-perception skills is influential in neuroscience. The fact that perceptual sensitivity to faces is found so widely in our species has led some to argue that humans have a hardwired face-detection capacity that is relatively immune to the influence of experience. The basic idea is this: Visual information from the world first gives rise to a basic representation in the brain of an object or event. As processing takes place up through the higher areas of the visual system, more content is added. At one stage (in a part of the brain known as area V4), color gets introduced to the visual representations; in another area (V5), motion is introduced. Objects are not visually categorized until they reach an even higher level of the extrastriate cortex (known as LO). An object is characterized as a face only if the object representation causes a particular *face-perception mechanism* to light up. This face-detection module, in turn, can be thought to face (so to speak) a computational challenge, namely, deciding whether the representation presented to it is a face or not. Face-recognition software is now a booming and, apparently, successful area of engineering. Face-recognition systems are being installed at airports as part of security screens. The very fact that such devices now seem technologically feasible suggests that we can get a grip on how the face module solves its computational task. The face module is the neural correlate of face consciousness.

Is there evidence to support this idea that perceiving faces is something the brain accomplishes in this way? There is no doubt, as we have already recognized, that faces are special: they are ubiquitous and they matter to us. Faces are special as perceptual objects in other ways too. For one thing, we exhibit an interest in faces from our earliest childhood. It has been shown, for example, that children in their first hours of life are drawn to faces; their interest is captured by faces even though their visual powers are very primitive at this stage and the face can show up for the child as, at most, a rough configuration. Another fact about face perception is also striking: We perceive faces holistically. We don't see faces by first seeing their individual features; rather, we take in the face somehow as a whole. That is why, as has been demonstrated, the swiftness and skill with which we can identify faces drops off sharply when faces are seen upside down. This "inversion effect" would be surprising if we identified a face by first identifying its features.

Scientists have actually discovered that there is a neural structure, or rather, a place, in the fusiform gyrus (now known as the fusiform face area, or FFA) that is strongly activated not only when subjects see faces but when they think about or imagine faces. The FFA would seem to be the place in the brain where faces are represented in consciousness. Further evidence for this is the fact that damage to the FFA would appear to produce face-specific visual deficits. This unusual disorder, prosopagnosia, is characterized not so much by an inability to pick out faces as by an inability to recognize individual faces.

Nancy Kanwisher, who has championed this proposal, argues that work on the face-recognition system lends support to the idea that the human brain is like a Swiss Army knife, "composed of special-purpose components, each tailored to solve a single specific task." The face-detection module, Kanwisher holds, is a domain-specific system for "representing the contents of conscious awareness."

Now, this innate face-detection module hypothesis—let's call it the Swiss Army knife theory—has been hugely influential in cognitive neuroscience; it stands as a kind of paradigm of what the new neuroscience of perception and consciousness can achieve. Is it plausible? I think not.

"The Familiar Face of a Word"

We can begin to approach this issue by considering as a comparison a different domain of perceptual expertise: reading. The process whereby children learn to read is painstaking and challenging. In the beginning of the process, children acquaint themselves with letters and learn the relation between the letters and sounds. They read words by reading each letter and sounding out the word. Once a reader has attained real fluency—say, after about age thirteen or so—things change. The mature reader has a relationship to words that is qualitatively distinct from that of the novice. In particular, there is striking evidence that mature readers are sensitive to words holistically. The amount of time it takes to read a word is not a function of the number of letters in it; a reader's ability to identify letters in a word is improved if the word is a bona fide word rather than a mere jumble. Finally, as we all know from experience, it is difficult to read upside-down text, at least without special practice. Researchers have shown that there is an area in the left fusiform gyrus (the FFA is normally in the right fusiform gyrus) that is responsive to the visual presence of words. This area lights up not only for words but for wordlike assemblages of letters (i.e., pseudowords). Finally, there is evidence that damage in this so-called visual word form area (VWFA) produces specific deficits in word perception.

I bring up this example because I hope it is now clear that we have no better reason to think that we have an innate face-detection mechanism than that we have an innate word-detection mechanism. But we know that we don't have an innate word-detection mechanism; after all, reading is cultural technology invented only relatively recently in human evolutionary history. So the view of the FFA advanced by Kanwisher is, at best, unsettled.

It is true that all normal people show expert facility with faces whereas only educated readers show word-recognition facility, so this may seem to make the case for a face-dedicated perception mechanism stronger. But a better explanation of this difference is the fact that all people are exposed to faces from infancy, whereas only some people are exposed to words. It is striking that if you are exposed to words, words become salient and attention grabbing in much the way that faces are. For example, you enter a room and there is an offensive bit of racist graffiti on the wall. You see it and the heat rises to your cheeks; your heart begins to flutter; you feel strong emotion. The words on the wall, together with their import and significance, simply pop out at you and, as it were, smack you in the face.

In fact, the claim that our sensitivity to faces is unaffected by experience is also not one that we can really take seriously. A simple example suffices: When I and a friend were traveling into Nigeria overland from neighboring Niger, the border guards asked us whether we were siblings, so strong, they insisted, was our resemblance. In fact, to someone from the same cultural milieu as ours, there was no resemblance between us whatsoever. Psychologists have shown that face-discrimination skills are sharply reduced when it comes to faces of "racial" groups other than one's own. We are better at differentiating the faces of the kinds of people we are most familiar with. This observation shows that at least some of our fine skill with faces is learned.

Faces are special, no doubt. So it isn't surprising that we can pick them out so confidently. But none of this adds up to a reason to endorse the idea that there is, inside us, a special inborn structure dedicated to the discrimination of faces. In fact, there is a superabundance of evidence to support the claim that, as in the case of words, our face-detection abilities are a special application of our more general ability to develop perceptual expertise. It turns out, for example, that the FFA is activated by nonfacial objects, but only when the objects belong to a class in which the perceiver has expertise. Birds light up the FFA in birders, cars light up the FFA in car buffs, and so on. This suggests that it is the fact that, as it happens, everyone is a face expert that explains why the FFA lights up in everyone for faces; only some of us are birders or car buffs, so only some of us show activation in the FFA for those different sorts of objects.

In a way, this makes a lot of intuitive sense. If you really know a domain, then you can pick out individuals in that domain; by contrast, novices are confined to cruder distinctions. Just as the primatologist sees a marmoset but not a tamarin where you or I see nothing more than a monkey, so all of us happen to be able to discriminate individual faces. Knowledge and experience enable you to experience the faces, if you like, of things like horses, cars, and monkeys, just as knowledge and experience enable us accurately to discriminate human faces.

It turns out, moreover, that the prosopagnosia data are far from conclusive. First of all, actual cases of truly face-specific visual deficits are extremely rare; usually prosopagnosiacs show other perceptual shortcomings. But more important, damage to the FFA in people who have expert knowledge of objects in a certain domain of nonfacial objects is likely to produce deficits precisely in the perceptual exercise of this knowledge.

My aim in this discussion of face perception is not to diminish the importance of the neural processes that make perceptual expertise possible. It would be incomprehensible if our acquaintance with objects and our development of expert powers of perceptual discrimination—with faces but also with words, letters, cars, birds, paintings, etc.—did not incorporate changes in our nervous system. What we can reject outright, though—we should know better—is that we can make intelligible the idea that a particular bit of brain tissue (such as the FFA) can really be the source of faces in our consciousness. To understand the FFA and why it behaves as it does, we need to keep clearly in mind the role that faces and other objects of perceptual awareness play in our lives. We cannot explain human perceptual expertise in terms of the FFA; nonetheless, the FFA figures in a fuller understanding of perceptual learning.

Bad Habits

Smoking, watching TV, chewing with your mouth full, interrupting other people, are considered *habits*; they are said to be *bad* habits. They are thought to be bad in part because they are habits. The idea here, I think, is this: If we are at our best when we are deliberative, then we are at our worst when we are habitual. Habitual action is thoughtless and uncontrolled; it is, in that sense, involuntary and brute. This suggests a further reason why habits are to be deplored. Insofar as we act from habit, then our actions are transparently revealing—that is to say, they are exposing. To be exposed is to be understood. And to be understood (as my father likes to say) is to be found out.

Many habits are in this way unseemly. This sort of personal

habit is made to be broken, and I suspect that there is something good, something healthy, about shaking things up by disrupting one's habitual regimes. But I entirely reject the idea that it would be best to be entirely habit-free (as if that were even possible!). I do not mean merely that there are in fact good habits. I mean rather that a habit-free existence would be a robotic existence.

Let me spell this out. First, you need habits of thought and behavior in order to be decisive and deliberate, for habit is the foundation of skill. This claim is obvious in sports and music, where training—repetition and drill—is the concrete foundation on which the structure of play gets erected. But this is no less true in obviously intellectual endeavors such as mathematics. We memorize basic elements of arithmetic (e.g., how to count, addition and times tables, techniques of calculation), and we use these skills to leverage ourselves up to ledges of insight we couldn't reach otherwise. It is true, then, that science and learning require foundations. But the true foundation is not true belief or necessary truth. At the foundation of science is basic, acquired, practical skill.

Whereas a person automatically looks at what you are pointing at when you offer an explanation, a dog or a cat more likely looks at your hand. Wittgenstein noticed that this natural capacity we have to attend jointly to an object of interest is an unlearned primitive skill without which communication probably would not be possible. Intellectualists believe that the human propensity for judgment and rational discourse is our distinctive cognitive accomplishment. I would say that this sort of cognitive virtuosity is a late addition, a fruit borne of the tree of practical skill. And practical skill itself is, at least at its basic levels, independent of our possession of the intellectual powers that depend on them.

Second, habitual modes of thought and behavior are themselves frequently expressions of intelligence and understanding, even if they are spontaneous, automatic responses to things. Indeed, they are expressions of understanding and intelligence precisely because they are habitual. Insofar as habits comprise those basic skills whose possession allows us to be expert walkers, talkers, readers, cooks, parents, and drivers, then, to just that extent, those basic habits of movement and management are themselves loaded with cognitive achievement. Take even so unappealing a habit as the tendency to clear one's throat before speaking. A habit such as this is probably deployed, however rudely, in a way that signifies one's grip on the basic rhythm and cycle of conversation. And so the habit itself signals, for one and all, one's immediate awareness of where one is in a conversational space.

Third, a life without habits would be robotic. Each day would be like one's first day in an unfamiliar country. No familiar pathways or tested strategies for getting things done would be available; no routines would be in place to serve as an anchor. Nothing could be taken for granted. We would scan, interpret, evaluate, decide, execute, reevaluate. Crucially, our actual living bears no resemblance to such a robotically alienated mode of being. For we are always already in full stream. Just as the moves in chess worth considering are fixed by the situation, so, more generally, our presence in a familiar context reduces to a manageable number the choices and degrees of freedom with which we need to concern ourselves. We don't need to test the solidity of the floor when we get up out of bed in the morning, and we know where to find the light switch and the bathroom. Radical changes can happen overnight—the Twin Towers can be attacked, an earthquake can occur, a loved one can have a heart attack—but what is striking about such occurrences is that they can seem, at least momentarily, to have the power to destroy us, to overwhelm us. That they don't is testament to the secure network of life structures that keep us standing upright even in the face of the unthinkable.

Good Habits

I have been arguing that we should reject the intellectualist conception, even as an account of the intellect. To play chess is to be at home in the chess environment; to be a language user is to be engaged in the forms of life with which linguistic practice is inextricably bound. A basic level of skillful expertise is the stage setting that first makes it possible for us to be able to pause, and deliberate, and wonder. This view is beautifully illustrated by simple perceptual situations. Consider: I go into the art gallery and sit down before a puzzling sculpture. What is it? Ah, of course, it is a human form; now I see it! But wait: I look closer at the way the body is bent double; this sculpture is clearly meant to serve as a chair. Now, contrast this contemplative, puzzled, interpretive attitude I take to the sculpture with the attitude I took to the bench I plunked myself down on to study better the work of art. Insofar as I perceived the bench and sat down on it, my actions gave expression to a basic understanding, familiarity, and perceptual and conceptual competence. But my basic perceptual sensitivity to what the bench is—namely, that it is a bench on which I can now rest myself while I contemplate the exhibit—was not something that itself required deliberation on my part. Indeed, if I had had to stop and wonder whether in fact this was a bench, or whether this structure was meant for sitting, that would have showed that something was wrong, or at least problematic, in my relation to the world around me. (I am ignoring, for present purposes, the fact that an art gallery is just the place where our relation to ordinary things may be called into question deliberately—i.e., that the "bench" may actually be another work of art on display.) In normal, everyday cases, the familiar world just shows up for us as what it is. The bench showed up as a veritable invitation to sit down and not as something that I needed to look at, categorize, evaluate, and then, only then, make use of. Heidegger has a term for this way in which things can show up for us: much of the time, things appear for us not as objects but rather as "equipment" (in German, *Zeug*).

Cognizance of the world around us is made possible by our possession of skills. These skills—our expertise—enable a great deal of our doings and sayings to be unplanned, unmediated, undeliberated, but, for all that, expressions of understanding, sensitivity, appropriateness, and responsiveness to what is going on around us. We nod at people when we pass them and understand angry gestures for just that; we take seats and use glasses to drink and turn on the lights and climb stairs; and we do all this just as a matter of course. When one visits a foreign country— even one that is, comparatively speaking, closely related to one's own (the United States and the European countries, say)—one has the feeling of being slightly disjointed, out of balance and alienated, as if everything had been rotated about some indeterminate axis. Now light switches may not work quite the same way; toilets flush differently; elevator buttons turn up where one doesn't expect them; and it is difficult to look up a number in the phone book. The disruption that we feel when we travel is a giveaway of our usual unthinking reliance on that background of skills that make functioning in the world possible.

Trails

The natural world is one that is sculpted by the life processes that occur within it. Living beings alter their environment and so alter their landscape of possible actions; that they do this is no less inevitable than that they create waste. We can see this phenomenon, this interplay of organism and environment, at many different levels. Plants and insects are colorful because there are animals around that are sensitive to color; animals have evolved sensitivity to color in order better to make out what there is in the environment. The oxygen-rich atmosphere of Earth is not merely the background condition necessary for life: living plants have manufactured, or rather, photosynthesized, the very air that animals need to breathe. The river cuts a divide through the rock and so builds a contoured landscape. Even a geological process of this kind is affected by animal and vegetable life.

Catastrophe can strike from outside the system, to be sure. For example, a meteor can strike Earth and in one fell swoop change the environmental conditions that living beings must face. But in the normal course of events, the texture and quality of the environment are both a prerequisite for and a product of life as we know it.

The mutual interdependence of organism and environment is exemplified in the existence of the path or trail. Trails are made by the very act of walking: our movements pat down the earth and sweep aside rocks and vegetation. Once the trail has come to be, it is difficult to avoid using it. We travel along grooves that our own repeated action has made for us; the paths we take are well-worn because we take them every day, and we take them in part because, being so well-worn, they are the paths of least resistance and also because venturing off the beaten path demands more work, and even risk. Just as the trickle of water creates a groove that, once existent, will attract ever greater quantities of water to it, so our own locomotion changes the ground itself and constrains our subsequent actions.

Most of us live in cities. But what is a city but a highly fortified structure of well-beaten trails and paths of least resistance? Goethe said that architecture is frozen music. It would be more precise to say that architecture and the cities to which it gives rise are frozen habit. Our possible movements are now laid down once and for all in concrete. Of course, it is not as if cities really are frozen. Cities and buildings belong to our natural environment; a truly static city, from a human point of view, would be a dead or abandoned city. It is also worthwhile to avoid the romantic idea that there is freedom only outside the city. The hiker would be well advised to keep to the trails unless she is able to fend for herself.

Here is an exercise: Plot the course of your movements on a map for the next month. If you are like most people, then you will find at the end of the month that certain routes are blackened out through repetition; here and there, occasionally, a thin line will sprout off the thick rope of daily routine. Vast stretches of your hometown are never visited. So predictable is our adherence to familiar routes that it is almost as if, like the water rushing along in the bed of the stream, we have no choice at all. What fixes our course? Are we so unimaginative that we don't even consider altering our journey?

The riverbed of habit makes travel along certain lines safe and reliable, efficient and easy. Have you ever gone on vacation in a new city and found that, after the first day, your movements tend to be confined to certain tried and tested routes? You use the same train station and change your money at the same bank; you take breakfast in the same café. Trying something new is always risky; by relying on what has been tested, we save our energy for the excursions that count the most (e.g., the trip to the museum or the theater).

The Limits of the Known World

As with cities and transportation, so with thinking, reading, conversation, friendship, and politics. We travel along familiar paths of thought and intellectual exploration not because we are lazy but because we must. It is almost impossible to beat a path through thickets and brambles; if we want to go someplace and do something, we have to stick to the path. And so it is with our intellectual lives. The paths we lay down are paved with the skills needed to move forward. Chess players memorize openings and endgames so they can lend structure to the improvisation of the middle game. Scholars master what has been written and said already so that they know what would even count as original. Artists, writers, and filmmakers are no less constrained by what has gone before, which marks the spot where there is still work to be done in this genre or in that medium. Radical novelty is almost impossible, and if we were to stumble upon it, it would take unusual strength and power to see anything of value in it. The challenge faced by an artist is to make something new that is comprehensible; to be comprehensible it must already be, in a way, at least partly, old. In fact, this is the quandary we face in every aspect of our lives.

Jazz, as an improvisational musical form, nicely illustrates the basic structure we are examining. Every performance is unlike any other; yet the novelty and variation, the improvising, take place according to more or less explicitly agreed-upon rules. What is extraordinary about a good solo is that it manages to make a statement and do something surprising within a highly predictable and conventionalized scheme. Indeed, this is true even of far less improvisational forms of music.

My point is not that novelty is impossible (although, in fact, radical novelty is basically impossible). My point is that this fact explains how progress can even be possible. To "boldly go where no man [or woman] has gone before," we must first travel to the limits of the known world. We must master the skills and habits that form the groundwork of all animal life.

The Ecology of Habit

Habits are basic and foundational aspects of our mental lives. Without habit, there is no calculation, no speech, no thought, no recognition, no game playing. Only a creature with habits like ours could have anything like a mind like ours. But habits, at least many of them, are situational or environmental. A habit is like a trail laid down by our own repetitive action. A habit is not merely a disposition to act or an automatic or unthinking tendency; it is a responsiveness to the environment in which we find ourselves. If the trail is paved over or if familiar landmarks are removed, our habits can frequently be extinguished. In the previous chapter we discussed the way in which skillful use of tools and technology (including language) enables us to transform not only what we can do but also our own sense of ourselves. Likewise, to remove those external tools and structures would be to eliminate the mastery on which our manner of activity depends.

The case of language is illuminating. You might grant that we can do things intellectually that we would be unable to do without language—that linguistic skills are the substrate of a wealth of cognitive capacities (thought about numbers and other abstract objects, distant times and places, etc.). But that is a far cry from the idea that if language were somehow to be made to disappear, we would lose our intellectual capacities. And this is precisely because language cannot be made to disappear. Once we know language, we own it; it is internal to what we are. Even if every other person in the world and all the books and papers and the Internet were to disappear overnight, we would not lose our linguistic skills.

Insofar as I have acquired language, what happens to others doesn't have a direct radical effect on my own skills. But it would be a mistake to think that this shows that language, really, is in-

ternal. If you'd been born into an isolated existence—with no
contact with others—you would never develop normal cognitive
competence and you certainly would not develop language. But
consider also: it is in fact an empirical question as to what extent
our linguistic capacities would survive and sustain themselves
in the absence of any social, external linguistic resources. No
doubt our most basic linguistic skills would persist. But there
might very well be a striking deterioration, not so much in how
good we are at using language, but rather in what it is that we
can use language to do. To return to Putnam's example discussed
in Chapter 4, I can't tell beech trees and elms apart; in a world
in which all repositories of information about what the difference is between elms and beeches had disappeared, my attempts
to construct thoughts about elms or beeches would start to lose
their point; that is, this particular corner of language would go
dead for me. Sure, I'd remember what I used to say about elms,
or what other people used to say. But like a coin of a now disused
currency, the meaning would vacate the word; it would be just a
fossil of a former life. But so much of our linguistic use is in this
way buoyed by the availability of other people, and of books and
libraries, TV and movies, that enable us to use words to reach
out and refer to that which is beyond our experience. I suspect
that in the absence of those structures that normally enable language to hook onto the world in the manifold ways that it does,
language would start to die on the vine. A solitary individual in a
world without writing or recordings of others would manage to
keep language alive, but only in an atrophied form. And with the
atrophying of language would go also the deterioration of capacities for thought.

Another good comparison would be sports. My ability to play
baseball wouldn't vanish in the absence of other people and of
the normal equipment. But insofar as baseball is an activity that
is nourished by the contributions of others, by actual play, it

seems clear that even my cognitive relation to baseball would shrink in these circumstances.

We can think of language as a tool, but only if we appreciate what a subtle sort of tool it is. The dependence of our skills on the use of more straightforward tools, such as hammers and cars, is correspondingly more dramatic. To be deprived of a tool is comparable to undergoing an amputation. To lose the tool is, most of the time, to lose the habit and skill. Our habitual mode of being is dependent for its actualization and its sustenance on the availability of the right kind of environment.

Conclusion: We Are Creatures of Habit and Habit Is World-Involving

Insofar as we are skillful and expert, we are not deliberate in what we do. Our skill enables us to respond appropriately to the world and in an automatic way. If we were to deliberate, we would interrupt the flow and undermine the conditions of our own expertise. We would choke. An appreciation of habit and practical skill in our intellectual life reveals that intellectualism is a misguided conception of even our intellectual capacities. Habit and skills, however, are world-involving. Just as my habitual route to work is shaped in part by the landscape in which I find myself, so in general our habits are made possible by the world's being as it is (even if it is also true that our action shapes the world in turn). The idea that the brain alone can explain the character of our conscious lives appears in this way thinner and more far-fetched. Neural activity enables us to develop the forms of expertise that determine how we deal with the world around us, but the brain is never more than part of the story about how all this works.

The basic fact of animal consciousness—we think, and a world shows up for us—can be explained only by supposing that we possess the kinds of skills that yield access to the world. We are not in the predicament that has been supposed traditionally by cognitive science, namely, that of having to figure everything out from first principles. Our lives depend on what Adrian Cussins, a British philosopher now at the National University of Colombia in Bogotá, has called cognitive trails and other modes of cognitive habits that presuppose for their activation our actual presence in an environment hospitable to us.

Why do so many thinkers persist in thinking that consciousness—thought, perception, the fact that a world shows up for us—can be explained in terms of interior neural events alone? And is there really any case to be made in favor of this idea at all? We must confront this insistence. I turn to this in the next three chapters.

6

THE GRAND ILLUSION

Men are empty heads turned towards one single self-evident world.   —Maurice Merleau-Ponty

The idea that our perceptual consciousness is somehow a grand illusion has seemed to gain support from work in empirical science. This is one source of resistance to the idea that the brain is only one element in the more complicated dynamic that is our conscious life. In this chapter I look at two influential lines of argument for this skeptical conclusion and show that neither has any force. In fact, there is no empirical basis for the idea that the world is a grand illusion.

The Creator Brain

Some neuroscientists believe that the brain is the power that creates the world and that it does so according to its own conception. We ourselves are brains in vats; that is, we are brains in biologically evolved vats of skin and bone. And we are victims of an illusion on a grand scale, for when we see and touch and hear, we misguidedly take ourselves to be in contact with the way things are here and now before us.

Many scientists savor these contrarian discoveries and proclaim them confidently. What we are aware of is never more than a picture or model constructed by the brain according to its own rules. As Kandel, Schwartz, and Jessell write in their authoritative textbook *Essentials of Neural Science and Behavior*:

> The brain does not simply record the external world like a three-dimensional photograph. Rather, the brain constructs an internal representation of external physical events after first analyzing them into component parts. In scanning the visual field the brain simultaneously but separately analyzes the form of objects, their movement, and their color, all before putting together an image according to the brain's own rules. . . . [Therefore] the appearance of our perceptions as *direct* and *precise* images of the world is an illusion.

Chris Frith, a prominent British neuroscientist, echoes this same thought in a chapter of his book *Making Up the Mind*. The chapter is called "Our Perception of the World Is a Fantasy."

Are we fictional characters playing out roles in a narrative authored by our brain, as these scientists would have it? Is there actually empirical evidence to support this claim, as the neuroscientists cited above would seem to suggest?

Vision: A Case Study

When psychologists and neuroscientists seek to dazzle audiences with evidence to support the hypothesis that the world is a grand illusion (according to which the world is a figment created "for us" by our brains), they invariably turn to visual phenomena. Vision occupies a central place in the science of the mind. No other sensory or cognitive capacity is understood as well as vision.

Scientific theories of vision date back to antiquity, flourished in the Middle Ages, and bloomed with the birth of modern science. Many of the greatest names in the history of human learning have made vision their stalking horse: Plato, Aristotle, Euclid, Ptolemy, Alhazen, Galileo, Leonardo da Vinci, Kepler, Descartes, and Newton. In more recent years, as we'll discuss in the next chapter, the Nobel Prize in Medicine and Physiology has been awarded for research into the neurophysiology of the mammalian visual system. This enormous interest in vision may be the reflection of our natural bias. We are visual beings, after all; our most basic grasp of what things are—of what a tree is, for example, or of who Mama is—is visual.

It has even been argued that vision is unique among our senses. Whereas hearing enables us to make acquaintance with sounds produced by events and other goings-on around us—and whereas smell, for example, acquaints us with chemicals released in our immediate environment—only sight gives us objects and events themselves. When you hear an intruder break in, you don't directly hear him or her; you hear sounds he or she makes. But when you see the intruder, you see him or her directly; you do not see a mere image or visual appearance.

Whether this view of the uniqueness of vision is true or not, there is a consensus that a consideration of sense perception in general, and vision in particular, provides the strongest case in favor of the idea that the brain is the creator and that the world is a grand illusion. So let's take a short journey into a mainstream of thought about vision and the brain and ask ourselves: Does the nature of vision support the idea that the world is a figment conjured for us by our brains?

The Miracle of Sight

Vision science has for a few centuries now taken its start from the idea that what we see far exceeds what we receive in the form of sensory stimulation. There is a discrepancy between what we see and what we really get from the world around us in the form of reliable, informative sensory stimulation. The brain's job, it is proposed, is to make up for this discrepancy. The brain's job is to compensate for the impoverished visual stimulus.

This general style of argument—the appeal to the existence of discrepancies between the character of our seeing and the character of the information contained in the retinal image and the inference that it is the brain's task to make up for the lack— is ubiquitous in visual theory. Here are some leading examples of this reasoning at work.

The Inverted Retinal Image and the Cyclopean Character of Vision

After much speculation during the Middle Ages, Kepler sorted out the eye's optics, or the way in which light entering the eyes is bent and so ultimately brought to a focus. In particular, Kepler demonstrated that the retinal image is necessarily inverted: that the visible scene must, owing to the eye's optics, project an upside-down image on the interior of the eyeball. The question that would seem immediately to present itself—it certainly confounded Kepler—is this: How is it that we manage to see the world upright when our eye's image of the world is upside-down? The puzzle gets thornier when we consider further that there are two, not merely one, retinal images. And these two upside-down images are not identical. Why don't things show up visually in duplicate and a little blurry (as they do when you actually press your finger against your eyeball and force their cooperation to break down)?

Here we have a discrepancy between what we see (an upright, singular visual impression in sharp focus) and what is given (two discrepant upside-down images). Somehow, some way—so scientists have supposed since Kepler's day—the brain must make up for this difference by flipping the images around and combining them to form a single upright representation of the world corresponding to what we actually experience when we see. This reasoning—from discrepancy to what the brain must accomplish—is the foundation of almost all work in vision science.

The Resolving Power of the Eye Is Uneven

The resolving power of the eye is not uniform; there are many more rods and cones at the center of the eye (in the fovea) than at the periphery. In fact, the eye is able to create a focused image only at the focus of gaze. It's easy to demonstrate this. Look at a page of text you've never read before. Shut one eye and stare at a word at the center of the page. (This works with two eyes too.) Try as hard as you can *not* to move your eye. Without moving, you'll find that you are unable to make out more than a few words located where you are looking. In fact, it turns out there are very few color-sensitive receptors at the periphery of your eyes; this means that you are, it would seem, color-blind at the periphery of your visual field. Again, it's easy to demonstrate this. If you fix on a point straight ahead, you won't be able to tell whether a playing card placed a foot or so off to the left or right, or up or down, is red or black. Of course, we experience a visual scene that seems fully in focus and normally colored across its expanse, even though the data for this perception—the retinal image—contains no such focused and colorful representation. Conclusion and explanation: The brain must add in or fill in the color at the periphery; it builds the focus in to some internal picture of its own making.

The Retinal Image Is Unstable

Matters are made worse by the fact that the eyes move almost continuously. Several times a second they jitter and bounce; they also make saccades and micro-saccades—that is, sharp, ballistic movements. As a result, the projection of an object you perceive to be still in fact jumps around on your eyeball, and when you track a moving object, its image stays still on your retina while that of the stationary background races across your eyes. Again, in order to explain how we manage to experience a stable visual world, we need to suppose, it seems, that the stability is achieved at some later stage in the processing of the original retinal information.

The Blind Spot

In each eye there is a blind spot where there are no photoreceptors. And yet we experience no gap or discontinuity in the visual field. Close one eye and look at a uniform expanse of color—at the wall, say. Do you notice any discontinuity? There is a discontinuity in the retinal image of the wall, however. So your sense of the presence of a continuous world must be the result, or so the now familiar reasoning would have it, of the brain's active filling-in of the gappy image.

Obstructions

Veins crisscross the eyeball. Bits of organic material float freely in the eye itself. These materials obstruct and distort the passage of light on its way to the retina. Strangest of all, the retina itself is positioned backward; that is, the sensitive receptor itself is behind the web of nerve fibers that ultimately join to form the optic nerve. Light must wend its way through this morass of axons and dendrites. And yet none of this shows up in our visual experience—or, if it does, it does so only rarely, as when we experience "floaters."

The Third Dimension

Have you ever had the experience of noticing the sound of a lawn mower down the street, only to recognize that what you actually hear is a mosquito buzzing near your ear? The eye is vulnerable to this kind of mistake as well. A small object nearby can project the same pattern of retinal stimulation as a large object at a distance. All we are given, when we see, is the two-dimensional image. How can you discern size or distance from a two-dimensional projection? You can't: it's mathematically impossible. It would seem, then, that if we do in fact see spatial relations—if we can see size and distance—we don't do so directly. That information just isn't there in what is given to us.

Color

We noticed already that there are few color-sensitive photoreceptors (cones) at the periphery of the visual field. Nevertheless, we experience the visual scene as fully colored across its entire expanse. In addition, we experience objects as unchanging in their color even as lighting conditions change radically. For example, most of the time the color of your book does not appear to change even though the physical character of the composition of the light reaching your eyes from the book's surface changes radically when you come inside into artificial lighting after having been outdoors in the midday sun.

Time

It is well-known that the stars we appear to see in the night sky may no longer exist, and if they exist, they may no longer exist in the form or location in which they show up for us. This is because it takes time for the light from the stars to travel the enormous distances necessary for it to reach us. It is less well-known, but no less well established, that the process of vision gets started only when light from objects, distant or nearby, reaches the eyes. Light produces electrochemical changes in our sensory periphery that then propagate at relatively slow speeds along the fibers of the nervous system. It is only when the appropriate signals reach their ultimate destination in the brain that we see! For this reason we never see things when we think we do; vision itself is necessarily a kind of time travel, an access not to how things are but to how they were moments before.

I have tried to bring out the fact that these various examples exhibit a common structure. They each illustrate the way in which the character of our experience seems to "go beyond" what is given and thus directs us to precisely the place where there is work for the brain to do: it is the brain that supplies what the world leaves out. The main task of vision science, as it has been practiced since Kepler's day, is to explain the mechanisms whereby the brain enables us to see much more than is present in the retinal image. Somehow, on the basis of two, tiny, discrepant, distorted, jumpy, upside-down, gappy, unevenly resolved, only partially color-sensitive, time-delayed pictures in the eyes, we manage to enjoy a unified and stable scene of objects and properties spread out around us in full color and volume in three-dimensional space. The task of vision science, as it has been understood from the time of Kepler down to the present, is to understand how the brain accomplishes this seeming miracle.

You will notice, I am sure, that as this story is told, the world itself—the way things are beyond the scope of our awareness of them—just doesn't get into the act. According to this story, seeing happens "in here," somewhere between where the light strikes the eye and the back of the head. At best, it seems, the world gets into the story by serving as a kind of offstage background. The world causally perturbs the nervous system at its periphery (the senses), thus giving rise to the events that cause us to seem to see.

But all this apparent seeing of a world beyond the senses is just so much fantasy!

Pictures of the World in Mind

We have been considering the ways in which, it seems, the end product of the brain's visual activity is a rich, detailed image of the world. Scientists lay great emphasis on the richness of our seeing, on its detail and dazzle. The question of vision science boils down to explaining how we can enjoy uniformly detailed, high-resolution, brilliantly colored images of the world when really we see so very little.

Stage magicians and set designers have long understood that in many ways our visual experience is not as rich as it seems. The first principle of stagecraft has it that the hand is quicker than the eye. It turns out that, to a surprising extent, we see what we expect to see. For this reason we are very suggestible. If the magician gives us reason to think that he has taken the coin from one hand and placed it in the other, well, that's just what we are going to see. Seeing is believing because, in effect, believing is seeing!

Scientists have recently come around to these age-old insights. In a series of remarkable experiments and demonstrations—on what has come to be known as change blindness and inattentional blindness—it has been shown that we fail to see, or at least fail to notice that we see, a great deal of what is happening around us—unless, of course, that detail affects what is directly of interest or changes our sense of the overall gist of the scene.

Suppose some powerful demon intent on deceiving you mag-
ically replaced the person sitting across from you on the train
every time you blinked. Would you notice? You certainly think
you would! Most of the time, of course, when a change like this
occurs, the change itself attracts our attention to what is changing. We are very attuned to the little flickers or sudden movements that accompany abrupt changes. But what if the changes
and the associated attention-grabbing flickers occur when you
are not looking—for example, when you are blinking? Like the
old tease: I say, "Hey, what's that over there?" When you look, I
snatch a french fry. Unless you catch me in the act, would you
notice? Not if there was a heaping mound of fries on your plate,
so that the absence of one fry wouldn't make much of a difference. As for the vanishing train passengers: Would you notice
their disappearance if the actual switching occurred when you
were not looking, as with the french fries? You'd notice, maybe,
if the replacement was very radically unlike the person removed
(e.g., a young child in place of an old man). But a good deal of
the time, you wouldn't notice.

Or at least so recent work in perceptual psychology would
strongly suggest. Powerful demons are not readily available to
collaborate on this kind of experiment. But with some ingenuity
(and computers) scientists have managed to put this hypothesis
to the test. In a famous videotaped series of experiments, a
young college-age man stops an older, professorial-looking gen-
tleman in the middle of campus and asks for directions. The
young man shows the professor a map indicating where he
needs to go. While the professor is giving directions, a group of
workers carrying a door march by, passing between the professor
and the student. The professor continues with his directions and
both men go their separate ways. It turns out that the student
and the workers were confederates in this experiment and that
one of the workers took the student's place when he was ob-

scured momentarily by the door. The professor didn't notice that he finished up his conversation with a person different from the one with whom he began it. (This experiment was performed by Dan Simons and his group, then at Harvard University, in the late nineties.)

Other examples—this first one is also from Simons's group. You watch a video of kids tossing a ball around; you've been given the task of counting the number of times two kids in particular exchange the ball. When you are done you are informed that a person in a gorilla costume had danced across the image on the screen; you'd simply failed to see it because your attention was directed elsewhere. When the video is replayed, you guffaw with laughter and astonishment, incredulous that you could have missed the gorilla. Another example: You are given the task of arranging colored blocks to match a pattern depicted on a computer screen. Every time you shift your gaze from the monitor to the blocks in order to set one down in the right place, the arrangement on the screen changes. You find the task difficult but it takes a long time before you realize that the model is changing.

What do these phenomena—change blindness and inattentional blindness—tell us about ourselves? A number of thinkers have suggested that these studies provide a novel kind of evidence that the visual world is a grand illusion. Traditional defenders of this idea emphasize the claim that the brain builds up our internal picture of the world: what we experience is an internal picture confabulated by the brain, not the world itself. The line of argument I now have in mind suggests an even more disturbing, and more radical, skeptical thesis. According to this new skepticism, it isn't really true that the brain builds up an internal model of the world; the fact that it seems to us as if the brain builds up an internal model shows that we are even more profoundly deluded about the nature of our own experience. We think we experience all the world in sharp focus and uniform detail in our visual experience. In fact, we do not. If the old skepticism argued that we see much more than is given to us, the new skepticism presses instead the idea that we don't see more than is given to us, but we misguidedly think we do.

From the standpoint of the new skepticism, vision science takes a new shape. The old theory had it that the topic of vision science is understanding how the brain builds up an internal picture. The subject of the new vision science is explaining why it seems to us as if the brain does this, when in fact it does not.

"The World Is Its Own Model"

The new skepticism has been ably defended by Daniel Dennett, Susan Blackmore, and others. But it makes a mistake from the outset. It does not seem to us perceivers as if the brain builds up an internal model of the world; rather, it seems to us as if the world is here and we are here in it. When I look out the window, it doesn't seem to me as if all the environmental detail is represented in my consciousness; rather, the detail seems to me to be there, in the garden, past the fence, across the street. If I want to describe what I see, I turn my attention not to my internal model but to the world. What I see is never the content of a mental snapshot; the world does not seem to be reproduced inside me. Rather—and this is the key—the world seems available to me. What guarantees its availability is, first of all, its actually being here, and second, my possessing the skills needed to gain access to it. I gather the detail as I need it by turning my head or shifting my attention. Granted, I do have a sense now that the entire scene is present: it doesn't seem to me as if the scene is brought into being by the fact that now I am looking at it. But what explains this is that although I don't now represent all the visible detail at once, I do have *access* to all the detail—and moreover, in some basic, practical way, I know that I do. For example, when I look at a tomato on the counter before me, in what does my sense that the tomato has a back side consist? Just in the fact that I understand, in a practical, bodily way, that moving my eyes and head in relation to the tomato brings the tomato's reverse side into view.

So vision science—even the new, radically skeptical vision science—has been barking up the wrong tree. There is no need to explain how the brain makes it the case that we have all the detail in consciousness at once, because we don't. And it doesn't even seem to us as if we do! By the same token, if it should turn out that the brain does not enable us to represent all the detail in consciousness all at once, then it would not turn out that we were victims of a grand illusion. After all, to repeat: The world doesn't show up for me as present all at once in my mind. It shows up as within reach, as more or less nearby, as *more or less present.*

Change blindness doesn't show that we fail to experience the rich array of detail we seem to see. It shows something else: that our ability to sustain perceptual contact with the environment over time is not just a matter of there somehow being a picture of the scene in our brains; rather, it is a matter of access. And this, in turn, is a matter of skill. For example, seeing requires a practical understanding of the ways that moving one's eyes and one's head and one's body changes one's relation to what is going on around one. And also, significantly, it requires that we do not occupy a demon world like the one we considered in the previous section. Our ability to lock onto and keep track of the world— our sturdy perceptual access to the world—depends not only on our skills but also on the fact that the environment in which we find ourselves is governed by certain causal and physical regularities. Our perceptual consciousness of the world as a causally, spatially, temporally well-ordered, regular, and predictable place depends on the world's actually being that way.

And this situation should not be in the least surprising. After all, our perceptual consciousness is a biologically evolved capacity, and evolution takes place in a given environmental niche. Our perceptual skills have evolved for life on earth, not life in an environment in which objects materialize and vanish at the whim of supernatural deceivers (or engineers). So the fact that we are vulnerable to deception—in the psychology lab or at the movies—just reveals the context-bound performance limitations of our cognitive powers. It does not show that our cognitive powers are radically deluded!

It is no illusion that the world shows up for us in robust and variegated detail, although it is untrue that we enjoy detailed, stable internal depictions of that external world. But neither does it seem to us as if we do. It is the world that is detailed, and our perceptual consciousness of it depends not only on one's brain but on body-dependent skills—and on the world itself.

We are, to use Merleau-Ponty's phrase again, empty heads turned to the world. The world is not a construction of the brain, nor is it a product of our own conscious efforts. It is there for us; we are here in it. The conscious mind is not inside us; it is, it would be better to say, a kind of active attunement to the world, an achieved integration. It is the world itself, all around, that fixes the nature of conscious experience.

Back to Vision

Even though vision science hardly demonstrates that the visual world is a fabrication of the brain itself or that we are the victims of a grand illusion, many vision scientists have assumed just this. It is assumed, by more traditional vision scientists at least, that vision is an activity in which the brain builds up a representation of scenes corresponding to what we experience. This is the theoretical organizing principle; this is the basic starting point. Moreover, it is assumed that the data for vision—its beginnings—are to be found in the eyeballs, on the retinas themselves. Seeing is a process, then, that unfolds between the eyes and the back of the head; the world itself is only an exterior something-we-know-not-what.

Once we give up this assumption, it becomes clear that the supposedly dazzling demonstrations of the brain's constructive workings fail to give any support whatsoever to the Creator Brain picture.

Take, for example, the problem of the inverted and twofold retinal image. Neither you nor your brain nor anyone else (except maybe your ophthalmologist) sees the retinal image, so there isn't really any reason to think that its orientation is even relevant to the question of how the world itself shows up for you. The retinal image is an image in a mathematical sense; it is a projection or a mapping. The retinal image is not an image in the sense of a picture—or if it is, this is entirely accidental. How it looks, or how it reads, plays no role in its performance of its neurophysiological job description. Once we appreciate that the retinal image isn't something that we see, we lose a grip even on what it means to say that it is upside-down. Upside-down, one must ask, relative to what? Who's to say what counts as upside-down in the head relative to the tasks faced by the nervous system?

No self-respecting scientist would admit to supposing that the retinal image is a picture that is somehow scrutinized by the mind's eye, as it were. For every scientist realizes the fallacy implicit in this supposition: we haven't explained vision, after all, until we explain how the mind's eye manages to "see" the retinal picture. By means of yet another image inside the mind's eye?

Descartes has already called attention to this fallacy of supposing we see because there is something inside us that is seen. This has come to be known as the homunculus fallacy. But consider the following quote from Kandel, Schwartz, and Jessell's textbook: "The superior half of the visual field is projected onto the inferior (or ventral) half of the retina, and the inferior half of the visual field is projected onto the superior (or dorsal) half of the retina." Revealingly, they add, in parentheses, as if requiring no further comment: "The brain of course adjusts this inversion."

It is hard to see why one would even think that there is a problem or puzzle regarding the retinal image, let alone any need for the brain to "adjust the inversion," if one has really and truly jettisoned the misguided assumption that the retinal image is a picture.

So the problem of the retinal image isn't a real problem; it's a pseudoproblem, one that rests, ultimately, on the idea that the brain learns about the world by scrutinizing the retinal image. Ditto for the problem of the Cyclopean eye. Since we don't see *by* seeing retinal images, the fact that there would be two retinal images, or hundreds, makes no difference to what we see. Or rather, if it makes a difference, it makes a difference of a radically different kind than we are supposing when we wonder how we manage to see a unified world when we have two eyes.

Or consider the problem presented by the fact that we experience the visual world as sharply focused and uniformly detailed even though the eye, by dint of its nonuniform resolving power, cannot create such a representation, at least not without the brain's help. Again, we don't experience the retinal image; we don't experience *any* image, in that sense. We experience *the world*. And we do so not by depicting it internally but by securing access. What is true, certainly, is that if you fix on a point or on a single thing, then you can't make out what is on the periphery. But to say that you can't make out what is on the periphery is not to say that you have a blurry experience of the periphery, or that the periphery shows up for you as blank. What we can learn from the limits of what can be made out in a single fixation is that fixations are not the units for seeing; that is, we don't see, as it were, by combining fixations together as we might string together frames to make a moving image. We are not confined to what we can make out in a fixation unless we are at the eye doctor's office. Seeing is active. When you go to the theater or a baseball game, you sit up and look around and move your eyes and your head; in this way you engage with the event in front of you. (Indeed, even when you try to be still, your eyes move on their own, making saccades three or four times a second.) Seeing is a kind of coupling with the environment, one that requires attention, energy, and, most of the time, movement.

And so it is for all the other phenomena that supposedly show that visual experiences are made by the brain. Take the would-be problem of visual stability. Why would we suppose that the movement of the eyes (or the retinal image) amounts to a defect in the visual stimulus unless we take for granted that the way the brain represents motion in the world is by means of a code according to which movements of the retinal image correspond to movements of things around us. What evidence is there to think that the brain makes use of such a code? Simply stated: none.

The theory of vision starts, as we have noted, from the assumption that the task for vision science is to explain how the brain transforms the retinal image into the percept. And that assumption is tantamount to the view that vision is an internal process, in this sense like digestion. Far from supporting this conclusion, the various puzzles are themselves, one and all, artifacts of this starting assumption. When we give up the assumption, we lose the feeling of puzzlement.

Doing Without the Creation Myth

There is something primitive about the idea of a Creator Brain. It captures the imagination of scientists. In a similar way, many people find it natural to think of a divine creator of the natural world. In this chapter I've tried to show that there is no empirical evidence in support of the former idea. It is an unargued-for starting assumption. And I've also suggested that we can do without it. When we do, a host of puzzles about how we see get solved, which is just to say, we realize that they don't require solution, for they rest on mistaken premises.

An alternative conception of ourselves and the world around us can now take shape. I began to present this alternative in Chapter 3. Seeing is an activity of exploring the world, one that depends on the world and on the full character of our embodiment. Far from its being the case that the world is a grand illusion, we find that we are at home in the world, that we are of it. Perceptual consciousness arises from our entanglement with it.

Science cannot prove that we are victims of a grand illusion. One reason for this conclusion is that science itself is an intellectual project carried out by human beings. It is a mode of reflection on one and the same reality that we all know in daily life. Science can unmask this or that way in which what we believe about the world we live in on the basis of common sense is misguided. For example, science can teach us that tissue is made up of cells and that the sun does not literally rise in the sky even though it looks as though it does. But there is nothing in what science can teach that can show that perceptual consciousness is not a mode of encounter with the world around us. For the possibility of that very encounter is presupposed by the scientist him- or herself.

Conclusion: Giving Up the Grand Illusion

The "grand illusion" hypothesis is bad philosophy; the cognitive science that supposedly provides evidence in its favor is bad science. Excellent work in perceptual psychology—for example, work on change blindness—properly construed, in fact provides excellent reasons to think of ourselves not as victims of a grand illusion but, rather, as open to an environment that matters to us.

7

VOYAGES OF DISCOVERY

Neurophysiologists, most of them, are still under the influence of dualism, however much they deny philosophizing. They still assume that the brain is the seat of the mind. To say, in modern parlance, that it is a computer with a program, either inherited or acquired, that plans a voluntary action and then commands the muscles to move is only a little better than Descartes's theory, for to say this is still to remain confined within the doctrine of responses.

—J. J. Gibson

In this chapter I tell the story of Hubel and Wiesel's Nobel Prize–winning research into vision in mammals. The work rests, I show, on an untenable conception of vision and other mental powers as computational processes taking place in the brain. The main problem with the computational theory of mind is that it supposes, mistakenly, that mind arises out of events in the head. The legacy of Hubel and Wiesel's research must be called into question.

The Visual Brain in Action

In 1981, David Hubel and Torsten Wiesel were awarded the Nobel Prize for research on the neurophysiology of vision, research they had conducted at Johns Hopkins, and then at Harvard, from the late 1950s until about 1980. Hubel and Wiesel's research, and the fact that it earned them the highest acclaim from the scientific establishment, is an important landmark in the science of consciousness. Seeing, after all, is in the first instance a mode of animal consciousness. In fact, for humans at least, seeing plays a colossal role in our conscious lives. The world is open to our visual inspection and we rely on seeing to get what we want and find our way around. But more than that, for us human animals our world is a visual world. It is a world full of profiles and colors and vistas. The visual character of objects shapes how we conceive of them: we think of them as having, for example, fronts and backs and hidden aspects. Think how difficult it would be to frame an understanding of what is going on around us—or at least to frame an understanding like that which we enjoy—if we couldn't see.

It is sometimes said that we know more about how the brain enables us to see than we do about any other mental function of the brain. When people say this they usually have the work of Hubel and Wiesel in mind. For most of history it has been impossible to study the workings of the brain of a living human being or animal. How would you do it? The brain is not open to view; it is hidden in a case of bone. And even if it were not— even if the skull were transparent—the brain's functional organization is obscure and its complexity makes it explanatorily opaque.

Hubel and Wiesel's importance is that they seemed to find a way to exhibit the brain's workings in a way that made what it was doing intelligible: how the brain might be achieving the function of making us visually conscious. To an astonishing degree their work was and remains the standard by which the field measures itself. As I will now explain, the shortcomings of Hubel and Wiesel's approach remain, even now, the shortcomings of the field of research into the neural basis of consciousness.

The Basic Project

Hubel and Wiesel's story, the journey they took, is fascinating and instructive. Let's begin at the beginning. First, Hubel and Wiesel poked fine microelectrodes into the visual cortices of cats and monkeys in order to record the electrical behavior of single cells. Doing so inflicted damage on the animals—after all, the electrode had to cut a path through tissue. But damage was isolated and this procedure seemed to allow, at least for a time, the investigation of the more or less normal behavior of individual cells.

Hubel and Wiesel were not the first to record in this way from the cortex; Vernon Mountcastle, at Johns Hopkins University, had recorded from the somatosensory cortex. Others, notably Sir John Eccles, the great Australian physiologist, had pioneered the technique of single-unit (cell) recordings in the spinal cord. One contemporary of Hubel and Wiesel's, Jerome Lettvin, at MIT, remarked that he rejoiced that Eccles had rescued neurophysiology from "Sherringtonian ooze." What Lettvin seems to have meant was that it was Eccles who had first found a way to translate the macroscopic neurophysiology of Sir John Scott Sherrington to the microscopic level.

Stephen Kuffler, who was Hubel and Wiesel's mentor at Johns Hopkins, and also Horace Barlow, at Cambridge University, who was their rough contemporary, had by the mid-1950s made important discoveries about the behavior of retinal cells. The receptive field for a visual cell is the area of the retina whose stimulation causes the cell to alter its firing rate. (Alternatively, you can think of the cell's receptive field as the region in space in front of the animal to which a cell is responsive.) Kuffler found that retinal ganglion cells had receptive fields consisting, in effect, of concentric circles. For such "on-center" cells, a spot falling in the central region of the receptive field would activate the cell, whereas a ring-shaped annulus falling outside the central region would inhibit the cell's firing. A diffuse light falling equally over the whole receptive field would produce a weaker reaction than a spot falling only on the center. Off-center cells had it the other way around.

Hubel and Wiesel were impressed with Kuffler's discovery, and from the start it was clear what they would try to do: "The strategy . . . seemed obvious," Hubel wrote. "Torsten and I would simply extend Stephen Kuffler's work to the brain. We would record from geniculate cells and cortical cells, map receptive fields with small spots, and look for any further processing of visual information." Others had tried to do this, but without any significant success. The problem, it turned out, was figuring out what sorts of stimuli would produce activation in cortical cells; in other words, the problem was getting the cortical cells to respond at all. "The cells simply would not respond to our spots and annuli," Hubel lamented. They did crack the problem eventually; Hubel and Wiesel were the first scientists who figured out how to make cells in the visual cortex talk, as it is sometimes put.

Their first discovery came about by accident. They were trying to find ways to stimulate a cortical cell by using slides to project spots onto a screen in front of the animal. No matter where they projected the spots, they couldn't get any response in the cell they were measuring from. Hubel explained:

Then gradually we began to elicit some vague and inconsistent responses by stimulating somewhere in the mid-periphery of the retina. We were inserting the glass slide with its black spot into the slot of the ophthalmoscope when suddenly over the audiomonitor the cell went off like a machine gun. After some fussing and fiddling we found out what was happening. The response had nothing to do with the black dot. As the glass slide was inserted its edge was casting a faint but sharp shadow, a straight dark line on a light background. That was what the cell wanted, and it wanted it, moreover, in just one narrow range of orientations.

What Hubel and Wiesel had discovered was a cell whose response was triggered by lines at a certain orientation. After this initial discovery, progress was steady, if arduous. Hubel and Wiesel found classes of cells in the cat visual cortex whose receptive fields departed strikingly from those found in the retina or in the geniculate (a thalamic way station between the retina and the cortex). For example, they found cells that had receptive fields with a similar sort of antagonistic organization to that found in Kuffler's center/surround ganglion cells but without the circular symmetry characteristic of cells in the retina. The optimal stimuli for these cells were stationary lines and slits at particular positions and with fairly precise orientations. They called these cells "simple." They also found cells that were like simple cells insofar as they responded best to lines or edges of given orientations but, unlike simple cells, were insensitive to the position of the line within the receptive field. Hubel wrote that the behavior of these cells "can most easily be explained by supposing that the complex cells receive inputs from many simple cells, all of whose receptive fields have the same orientation but differ slightly in position." What Hubel and Wiesel came to believe is that the network of cells as a whole is organized hierarchically; that is to say, complex cells are driven by networks of simple cells.

This was just the beginning. Among the highlights of Hubel and Wiesel's nearly twenty-five-year research collaboration can be counted the finding that complex cells give an especially strong response when a line is swept across the field, with some cells firing better to one direction of movement than another, and also the discovery of even more specialized "hypercomplex" cells. These specialized cells were frequently said to be selective for orientations and directions of movement.

Hubel and Wiesel made significant strides in characterizing

the "functional architecture" of the visual cortex, finding, for example, that columns of cells with related receptive field properties formed functional units. Moreover, orientation columns were found to exist that are in effect, as Hubel put it, like "a little machine that takes care of contours in a certain orientation in a certain part of the visual field."

And then there are experiments on the consequences, for cortical development, of depriving newborn cats and monkeys of the use of their eyes by sewing their lids shut. Hubel and Wiesel demonstrated that deprivation during this critical period caused an irreversible lack of connectivity in the cortex resulting in permanent blindness. They showed that the ability to see requires experience. If prevented from seeing during a critical period, animals will never see.

Christopher Columbus and the Brain

Hubel and Wiesel's findings are impressive. These are hard facts, and they speak for themselves. That seems to have been Hubel and Wiesel's view. Hubel wrote: "Almost absent from our way of working and thinking were hypotheses, at least explicit ones. We regarded our work as mainly exploratory, and although some experiments were done to answer specific questions, most were done in the spirit of Columbus crossing the Atlantic to see what he would find." And he observed: "It is hard, now, to think back and realize just how free we were from any idea of what cortical cells might be doing in an animal's daily life."

What a strange and remarkable thing for Hubel to say! Christopher Columbus didn't just set off across the Atlantic to see what he would find. He had very precise and, as we now know, erroneous ideas about what he would find. But the storied explorer aside, it is impossible to take seriously the claim that

Hubel and Wiesel were not guided by theory and responsive to its demands. Indeed, how could they *not* have been? After all, there are billions of cells in the brain, and they are massively interconnected. To form any conception whatsoever of what individual cells are contributing to the brain's functioning, you need to have a tolerably clear idea in advance of what the brain is doing. And indeed, Hubel and Wiesel did have such a guiding conception.

In *Eye, Brain, and Vision*, published in 1995, Hubel wrote: "We know reasonably well what it [the visual cortex] is 'for,' which is to say that we know what its nerve cells are doing most of the time in a person's everyday life and roughly what it contributes to the analysis of visual information." And he went on to add: "This state of knowledge is quite recent, and I can well remember, in the 1950s, looking at a microscopic slide of visual cortex, showing the millions of cells packed like eggs in a crate, and wondering what they all could conceivably be doing, and whether one would ever be able to find out."

Back in 1958, when Hubel and Wiesel set off on their journey of discovery, no one knew how neurons functioned so as to contribute to the analysis of visual information. In that sense it is true that it was not known back then what the visual cortex is "for." But that the visual cortex was in the business of analyzing visual information (as Hubel put it) and that, therefore, individual neurons must somehow, some way, be making a contribution to that was something that Hubel and Wiesel knew from the moment they set sail. Or rather, it was something they took for granted and assumed. Consider again Hubel's remark, first cited in the last section, now with italics added: "The strategy... seemed obvious. Torsten and I would simply extend Stephen Kuffler's work to the brain. We would record from geniculate cells and cortical cells, map receptive fields with small spots, and *look for any further processing of visual information.*"

And they were not alone in taking an information-processing conception of the brain for granted. By the late 1950s it was common belief among neuroscientists that vision presented the brain with a problem in information processing, and that the parts of the brain dedicated to vision could be thought of as systems of networks or circuits or, as Hubel and Wiesel sometimes put it, machines for "transforming information" represented in one system of neurons into progressively more refined and complex representations of what is seen. For Hubel and Wiesel, the visual system consisted of cells whose receptive field properties made them, in effect, symbols for features such as edges, orientations, directions of movement, and color. Cells were understood to be specialized in order to be able to "stand for" and thus represent features. This application of information theory to the brain was not new when Hubel and Wiesel set to work. Santiago Ramón y Cajal's student Rafael Lorente de Nó had represented neural relations as networks already in the thirties, and his treatment had a direct influence on the work of Warren McCulloch, Walter Pitts, and, through them, John von Neumann. (Walter Freeman, the neuroscientist, likes to say that Lorente de Nó is in a way the godfather of the digital computer.) Interestingly, Claude Shannon, one of the developers of the mathematical theory of information, was skeptical that the brain was an information processor; he believed that processing information required a transmitter, a receiver, and an agreed-upon code, none of which is found in the brain. But Shannon's skepticism didn't do much to dampen the general enthusiasm for the new approach. And so, twenty-five years after the beginning of their joint collaboration, Hubel and Wiesel were awarded the Nobel Prize "for their discoveries concerning information processing in the visual system."

Hubel and Wiesel set out from the start to understand how the behavior of individual cells, and their organization into larger assemblies, could accomplish the information-analysis task that is vision. They took for granted that vision was a process of analysis of information. It is remarkable that their landmark investigations into the biology of vision take as their starting point a startlingly nonbiological engineering conception of what seeing is.

The Computer Model of the Mind

David Marr, whose landmark book *Vision* was published in 1982, the year after Hubel and Wiesel won the Nobel Prize, made explicit the conception of vision on which Hubel and Wiesel implicitly relied. Marr pronounced that vision is an information-analysis process carried out in the brain. This, of course, was how Hubel and Wiesel had understood vision all along. Vision is the process of discovering how things are in the scene from images in the eyes. That is, it is a process of extracting a representation of what is where in the scene from information about the character of light arrayed across the skin of receptors in the eyes.

As we have noted, the use of information-theoretic ideas to understand what the brain is doing was already well entrenched by the late fifties. As far back as the nineteenth century, Helmholtz had proposed that perception is inferential: the brain constructs and tests hypotheses about what sort of events in the world are producing these impressions. And indeed, as we saw in the last chapter, this has been a guiding idea in the study of vision over the course of the last century.

What was new in Marr's work—here he made an advance over Hubel and Wiesel—was the attention to and conceptual clarity regarding theory that he brought to the table. Marr wrote that "trying to understand vision by studying only neurons is like trying to understand bird flight by studying only feathers: It just cannot be done." You need a theoretical conception of what neurons (or feathers) are doing just in order to decide which facts are even relevant. That is, you need to characterize what the system is doing in a way that is more abstract. And this is not because of anything peculiar to vision but because of the particular explanatory challenges we face when we want to understand an information-processing mechanism.

Take a simple case. You can't understand how a particular cash register works if you don't understand what it is for, namely, adding up numbers to keep track of balances due. When it is clear that that's what the machine is doing, then you can reasonably ask: How does it manage to do this? And so you will have to investigate the different ways this machine, or any machine, might be organized so as to add up figures.

There are lots of different procedures or recipes for doing this—what mathematicians call algorithms. When you select an algorithm, you select, in effect, a way of representing a problem and a way of solving it. For example, we are all familiar with algorithms for doing addition that involve the use of a pencil, paper, and Arabic notation. You don't need to understand addition in any deep sense in order to add up numbers. Now, there are different algorithms for adding: you'd follow different procedures if you were representing numbers in Roman notation or in binary notation. Likewise, there are lots of different kinds of physical mechanisms that can be used to carry out the function of addition. You might use pencil and paper, or an abacus, or a mechanical cash register, or a digital computer. In your effort to understand how a given machine works, you need to grasp and answer three questions: First, what function is it computing? Second, what algorithms or rules is it using to carry out that function? Third, how are those algorithms implemented physically in the mechanism?

The beauty of this approach is that it enables the study of an information-processing mechanism to move forward even though the physics or electronics or physiology of the mechanism may be unknown. If vision is the process of producing a representation of the scene from information about the wavelength or intensity of points of light striking the eyes, then we can begin to investigate what kinds of rules would enable that kind of analysis of visual information even before we understand much about the behavior of cells in the eyes and the way they are linked together in networks. The information-processing approach to the mind, and to vision in particular, enables one to appreciate both that the processes are carried out in a physical medium (the brain, a computer, whatever) and that the processes are not themselves intrinsically physical. They are information-theoretic, or computational.

Again, we come up against the irony that vision is made amenable to neurophysiological investigation only at the price of conceptualizing vision as, in and of itself, a nonbiological (that is to say, a computational) process that just happens, in humans, to be realized in the brain. The fact that we can see thanks only to the workings of our wet, sticky, meat-slab brains doesn't make seeing an intrinsically neuronal activity any more than chess is. To understand how brains play chess, you first need to understand chess and the distinct problems it presents. And, crucially, you don't need to understand how brains work or how computers are electrically engineered to understand that. Chess is only played by systems (people and machines) made out of atoms and electrons. But chess isn't a phenomenon that can be understood at that level. And the same is so for vision. To understand how the brain functions to enable us to see, according to the information-processing perspective, you must understand vision as the sort of process that might unfold just as easily in a computer.

Is the Brain an Information Processor, Really?

Marr's idea, and Hubel and Wiesel's, is that the visual system—the vision-dedicated parts of the brain—performs an information-processing task: it extracts information about the environment from the retinal image, thus constructing an internal representation of that environment. For example, the brain notices sharp discontinuities in the intensity of light at different points; in its internal representation of the scene, it labels these places "edges." That's just what seeing is: a process whereby the brain takes patterns of light on the retina and transforms them into a representation of what is where in the scene before the eyes.

The information-processing approach to the brain and vision has been entrenched in science for almost a century. You might turn on the radio any day of the week and hear an author blandly state, as a matter of established fact, that language is "processed" in the left hemisphere of the brain or that it is the neocortex that computes higher cognitive functions. And we are not in the least nonplussed to learn that Marr, Hubel and Wiesel, and others hold that seeing is a neural process in which information is extracted by the visual system from the retinal image.

But is the brain an information processor, really? There is an obvious reason to question this conclusion. Consider: we know what it means to say that a detective, for example, extracts information about an intruder from a footprint, or that an oceanographer gathers information about a prehistoric climate by studying fossils of unicellular organisms that she dredges up from today's ocean floor. These are nice examples of "extracting information" about one thing from another. The explanation of the fact that the footprint and the fossils contain information about the intruder and the climate, respectively, is the further fact that there is a definite causal relationship between the character of the intruder and the properties of the footprints, or between the climate millions of years ago and the fossil chemistry of foraminifera today. And what makes it the case that the detective and the oceanographer can extract this information is that they are each armed with knowledge of the way in which what they have access to now (the footprint, the fossils) was shaped by what they want to learn.

Things are different, though, when it comes to the brain and the retinal image. No doubt the retinal image is rich in information about the scene before the eyes; after all, there are reliable and well-understood mechanisms whereby the former is brought into being by the latter. Presumably, then, a suitably placed scientist would be able to extract that information. But the brain is no scientist or detective; it doesn't know anything and it has no eyes to examine the retinal image. It has no capacity to make inferences about anything, let alone inferences about the remote environmental causes of the observable state of the retina. How, then, are we to make sense of the idea that the brain is an information-processing device?

. There is a danger of vacuity in this "computer model of the mind." Our goal is to understand the biological basis of mind. It is hard to see how we succeed in making a contribution to the attainment of this goal if we suppose that our own mental powers are to be explained with reference to the cognitive powers of the brain. We—adult humans and other animals—think; we see, we feel, we judge, we infer. It's working in a big, plain circle to say that what makes it possible for us to do all that—what explains these prodigious powers of mind—is the fact that our brains, like wily scientists, are able to figure out the distal causes of the retinal image. For that just takes for granted the nature of mental powers without explaining them. Is cognitive science guilty in this way of reasoning as if there were mind-possessing agencies (homunculi) at work inside us?

The Computational Brain

You may think that the existence of the digital computer—I
mean the ubiquitous, everyday consumer appliance—provides
proof positive that a mere mechanism such as the brain can
process information. Computers, after all, perform calculations;
they render three-dimensional models from line drawings. Computers correct spelling and play chess, and they do so, we know,
without magic or the use of little guys inside them. What better
grounds could there be for thinking we should take seriously the
thought that brains are, in effect, organic computers? Whatever
mystery might seem to attach to thinking of the brain as an infer-
ring, reasoning extractor of information shivers away when we
consider that even much simpler human-made artifacts like computers are capable in this way of thinking.

Some problems admit of mechanical solutions. If you want to
know how many people are in the room, you can count them.
Coming up with the solution requires no more than an ability to
add one again and again. Likewise, you don't need to understand
long division in order to find the answers to long-division problems. You just need to be careful. You were taught a decision
procedure in school, one that makes use of writing, the Arabic
system of notation, and the fact that you know how to divide, add,
subtract, and multiply very small numbers. Any sufficiently careful idiot can do it. A machine can do it. In this same way, you
don't need to be a great chef to follow a great chef's recipe, and
you don't need to be able to grasp the vast number of combina-
tions and permutations possible on a Rubik's Cube to learn the
tricks that allow you to "solve" it in seconds.

An algorithm is a recipe or procedure for solving a problem.
It is, as it were, a program that enables one (a child, an idiot, a
machine) to reach a desired conclusion in a finite number of
steps. Some problems can be solved by algorithms; some cannot.

There is no general procedure for deciding, for any given puzzle or problem, whether it is or is not "decidable" (as mathematicians say) by purely "mechanical," formal methods. It has been shown, however, that any problem that is mechanically (or "effectively") decidable can be computed by any of a class of formal systems. A digital computer, as we know it today, is one such physically realized formal system.

But it would be a mistake to think that these findings in the mathematics of computability—or that achievements in the domain of computer engineering—prove that our brains are, in effect, computers. For this claim is founded on a mistake. No computer actually performs a calculation, not even a simple one. Granted, following a recipe blindly and without comprehension is one way to find the answer to a puzzle. But, crucially, understanding a problem or a computation does not consist in merely following a rule blindly. As a little reflection on one's school days will make evident, there's all the difference in the world between understanding the solution to a problem and getting a good score on the test because you have memorized a recipe for doing so. Computers may generate an answer, but insofar as they do so by following rules blindly, they do so with no understanding.

But more important, computers don't even follow rules blindly. They don't follow recipes. Just as a wristwatch doesn't know what time it is even though we use it to keep track of the time, so the computer doesn't understand the operations that we perform with it. We think with computers, but computers don't think: they are tools. If computers are information processors, then they are information processors the way watches are. And that fact does not help us understand the powers of human cognition.

The Mind Is Not in the Head

Now, the fact that computers don't think may be a good reason
to hold that brains don't think *because* they are computers. The
philosopher John Searle, my colleague at UC Berkeley, has
made this point persuasively. This leads Searle to go on to claim
that consciousness and cognition arise from the intrinsic nature
of neural activity itself. They are "caused by and realized in the
human brain." Computers solve problems and represent the
world only derivatively, thanks to the fact that we treat them as if
they do. But the brain's powers are not derivative; they are original. The brain thinks and represents.

But this is exactly the wrong conclusion to draw from the fact
that brains don't think by computing: they don't, but not because they think some other way. Brains don't think. The idea that
a brain could represent the world on its own doesn't make any
more sense than the idea that mere marks on paper could signify
all on their own (that is, independently of the larger social practice of reading and writing). The world shows up for us thanks to
our interaction with it. It is not made in the brain or by the brain.
It is there for us and we have access to it. What makes it the case
that my thoughts are directed to this task (playing chess, say) or
to this object (a glass of water, for example) is not the intrinsic
nature of my internal computational states. I agree with Searle
on this score. But that's because what gives my thoughts their
content is my involvement with the world. On no account does
my interior makeup suffice all on its own to give meaning and
reference to my mental states. Meaning is not intrinsic, as the
philosopher Daniel Dennett has rightly argued; it is not internal.
Meaning is relational. And the relation itself thanks to which our
thoughts and ideas and images are directed to events, people,
and problems in the world is the fact of our being embedded
in and our dynamic interaction with the things around us. The
world is our ground; the world provides meaning.

The limitations of the computer model of the mind are the limitations of any approach to mind that restricts itself to the internal states of individuals. Cognitive science sought to reveal how the brain can be a subject of thought by supposing that it is, in effect, a kind of digital computer. But what now emerges is that computers cannot think (or see, or play chess), and for the very same reasons that brains can't.

The central claim of this book is that the brain is not, on its own, a source of experience or cognition. Experience and cognition are not bodily by-products. What gives the living animal's states their significance is the animal's dynamic engagement with the world around it.

The Mind-Body Problem for Robots

The movie *Blade Runner*, discussed briefly in Chapter 2, makes vivid what seems to be incontrovertible: there is no principled basis for denying the rebellious worker slaves the respect and consideration we believe to be required of us in relation to human beings. Certainly, no facts about what is going on inside replicants justify such a denial. Replicants are made, yes. But then, in a sense, so are we. And yes, they lack a certain autonomy; but then we, too, lack autonomy. Far into our adulthood, we depend on parents, family, friends, society, *just to survive*. Granted, replicants lack biological innards; they are not composed of the same stuff as we are. But that's just the point: there is no necessary connection between what we are and what we are made out of. It would be nothing but prejudice to insist that there is such a connection.

Of course there are all sorts of practical reasons to think that we need brains like ours to have minds like ours. The technology that would support artificial minds is a long way off. The point is that it is not as though we understand so well how we work that we can rule out, in advance of case-by-case consideration, whether we will one day discover, or even learn how to manufacture, different kinds of minds.

A curious upshot of this line of argument is that even if Searle is right that computers don't think and that, therefore, our brains don't think because they are computers, it remains an open empirical question whether we could build a conscious robot with a digital computer for a brain. And so it remains an open question whether our brains are, in some sense, computers.

Flaws in the Foundations

Marr, Hubel, and Wiesel may have been right that we can gain some insight into how the brain works by thinking of the brain as an information processor. That is, it is reasonable to take up a functional approach to the brain by asking ourselves: Just what sorts of problems is the brain solving? What is it doing? For this reason it is reasonable, as methodology, to approach seeing and other mental powers as information-processing capacities. But these authors seem to have entirely overlooked that this methodological decision on its own did not force them to think of vision as an information-processing problem carried out *in the brain.* Nothing forced the assumption that vision is a process that takes place between the eyeball and the back of the head. And it is that assumption—that the resources for understanding vision, however characterized, are internal to the brain—that dooms the approach. Neural activity simply cannot rise to the level of consciousness, not even when that neural activity is described in computer-theoretic terms.

Hubel and Wiesel's approach to vision is based on the idea that the brain sees by processing special kinds of signals or symbols; the brain sees by building up an internal picture. But the brain doesn't see; there is no reason to think that seeing happens in the brain. And what good are symbols if no one is around to read them?

Where other scientists were incautious—some trumpeted the existence of "grandmother neurons," i.e., neurons selective for very specific meaningful stimuli like the face of one's grand-mother—Hubel and Wiesel were anything but. They hesitated even to refer to their line- and orientation-selective cells as edge or orientation detectors, although that seems to be precisely how they thought of them. Consider this ambivalent passage from Hubel's Nobel lecture:

> Orientation-specific simple or complex cells "detect" or are specific for the direction of a short line segment. The cells are thus best not thought of as "line detectors": they are no more line detectors than they are curve detectors. If our perception of a certain line or curve depends on simple or complex cells it presumably depends on a whole set of them, and how the information from such sets of cells is assembled at subsequent stages in the path, to build up what we call "percepts" of lines or curves (if indeed anything like that happens at all), is still a complete mystery.

In this remarkable passage Hubel expresses deep and I think warranted concern about the very theoretical framework that in fact motivates and makes sense of their prizewinning research. If the visual system does not build up a "percept," to use his phrase, on the basis of the sort of processing that their work identifies, then we can be forgiven for asking: Why is it even relevant to an understanding of vision that there are "specialized" cells in the cortex that respond to specific kinds of stimuli of the sort that Hubel and Wiesel discovered? On the assumption that the visual cortex does build up a representation of the scene on the basis of information in the retina, then the existence of stimulus-selective cells might seem to be tantalizing evidence of how the computational process unfolds. But if it is a "complete mystery" how and whether the brain performs this computational task, as Hubel acknowledges, then it must be admitted that we really have no reason to think Hubel and Wiesel's discoveries tell us anything at all about the brain basis of vision.

This is a harsh conclusion, but one that is hard to avoid. The whole idea that signals are passed from receptors to retinal ganglion cells, up to geniculate cells, and then on to simple, complex and hypercomplex cells, eventually activating the experience of seeing the world, can and should be called into question. As discussed in an earlier chapter, we now know that there are more connections down from "higher" visual areas to lower visual areas than there are connections moving in the opposite direction. That is to say, there is feedback. So whatever is going on, it isn't the sort of simple hierarchical process that Hubel and Wiesel invoke.

We now know that the behavior of cells in the cortex varies, depending on what the animal is doing or what it is paying attention to. The modulation of the behavior of cells depending on the context of the animal's activity is something that Hubel and Wiesel's research did not and could not take into account, for they worked only with animals that were not engaged in any active task: their subjects were unconscious. That is, they were anesthetized, paralyzed, on artificial respiration; stimuli were presented to eyes whose lids were peeled back and held open with clips; eyes were kept moist and clear by means of contact lenses. It is only the assumption that vision is something that happens passively inside the brain that could justify conducting research of this sort in an effort to understand how vision works. But surely we should question this assumption. Remember, we have no clue how neural activations would or could make visual experience happen. Moreover, it is salutary to remember that animals evolved vision not to represent the world in the head but to enable engaged living—for example, the pursuit of prey and mates and the avoidance of predators and other dangers.

Conclusion: Mind Is Not the Brain's Software

Computers can't think on their own any more than hammers can pound in nails on their own. They are tools we use to think with. For this reason, we make no progress in trying to understand how brains think by supposing that they are computers. In any case, brains don't think: they don't have minds; animals do. To understand the contribution of the brain to the life of the mind, we need to give up once and for all the idea that our minds are achieved inside us by internal goings-on. Once this is clear, we are forced to rethink the value even of Nobel Prize–winning research. This is a disturbing consequence, but one we had better be willing to accept if we want to move forward with a genuinely biological theory of ourselves.

8

A NOTHING RESERVED
FOR EVERYTHING

Once in a while, I get shocked into upper wakefulness, I
turn a corner, see the ocean, and my heart tips over with
happiness—it feels so free! Then I have the idea that, as
well as beholding, I can also be beheld from yonder and am
not a discrete object but incorporated with the rest, with
universal sapphire, purplish blue. For what is this sea, this
atmosphere, doing within the eight-inch diameter of your
skull? (I say nothing of the sun and the galaxy which are also
there.) At the center of the beholder there must be space
for the whole, and this nothing-space is not an empty nothing but a nothing reserved for everything.

— Saul Bellow, *Humboldt's Gift*

Consciousness does not happen in our brains; it is not a product of the brain. Certainly there is no sound empirical evidence
to support the idea that the brain alone is enough for consciousness. But is there some more general reason to hold, as so many
neuroscientists do, that the brain alone is sufficient for human
consciousness? I turn to this issue now.

The Foundation Argument

In 1996, Francis Crick, then of the Salk Institute, wrote that "it
is not impossible that, with a little luck, we may glimpse the out-

line of a solution [to the scientific problem of consciousness] before the end of the century." He had the last century in mind, of
course. One of the main claims of this book is that we will remain lost in the woods of theorizing about consciousness if we
don't manage to call into question the assumptions about the nature of consciousness that mainstream neuroscience has been
taking for granted. It's time to address this issue in more detail.

So let us ask: Do we in fact have any reasons for taking seriously the idea that the brain alone makes us conscious? We can
think of this dogma as a Cartesian inheritance of establishment
neuroscience. Now, Descartes didn't believe that the brain is the
thinking thing (the *res cogitans*) within us that is, as it were, the
self we are. But from the perspective I'm developing in this
book, that's just a technical detail. Establishment neuroscience is
committed to the Cartesian doctrine that there is a thing within
us that thinks and feels. Where the neuroscientific establishment
breaks with Descartes is in supposing that that thinking thing is
the brain.

Are there reasons why we should think that the brain alone
makes us conscious, or that consciousness (thought, feeling, perception) arises inside us, in our brains? We spend all our lives
embodied and situated and involved with the world around us.
How could we take seriously the idea that consciousness depends only on what happens inside the brain? This idea may be
good enough for science fiction, but why should we let it frame
the way we understand ourselves?

This question gets sharper still when we realize that our
brains, no less than our hearts and hands, evolved in a particular
environmental situation and under selection pressures appropriate to that situation. Indeed, our individual development, in
utero and after birth, is shaped by environmental and evolutionary pressures. Again, we can ask: Where would we even get the
idea that the brain alone is enough for our conscious lives?

My position is simple: Cartesian neuroscience has no empirical support for its basic assumption that conscious experience is an exhaustively neural phenomenon. So we need to look elsewhere to find the foundations of this unquestioned commitment. What we find is that, when pressed, scientists are likely to appeal to what is, in fact, a bit of traditional philosophical argumentation. It goes like this: The fact that we dream, and that we can produce events in consciousness by directly stimulating the brain, shows that the brain alone is sufficient for consciousness. Let's call this the Foundation Argument. It is simple and it can seem compelling. In this chapter I will explain why the Foundation Argument is not convincing.

Weak Foundations

According to the Foundation Argument, the fact that we can produce experience by direct stimulation of the brain shows that it is the brain itself, independent of its larger context, that is the basic ground of experience. But this is entirely unsatisfactory. First, it just isn't true that we know how to produce experience by directly stimulating the brain. Yes, we can produce some events in consciousness this way. For example, electrodes placed appropriately in the living brain of a conscious person can give rise to sensations of light (known as phosphenes). And given that we are already visually aware, then appropriately placed zaps of magnetic pulses (transcranial magnetic stimulation) can modify the quality of our seeing. For example, illusions of motion can be introduced by zapping cells in the middle temporal area of the brain (known as the MT). But from the fact that we can produce some experiences by direct action on the brain, it does not follow that we can produce all events in consciousness in this way. To suppose otherwise is simply to make a mistake.

Second, suppose that it were possible, thanks to some not-yet-invented or even imagined technologies, to create elaborate hallucinations corresponding to all our normal sensory experiences. That would not show that the brain alone is sufficient for these hallucinatory experiences, let alone for all consciousness. The best we can say is that it would show that the brain plus the actions of the manipulating scientist are sufficient for the occurrence of hallucinatory events in consciousness. What we imagine, in this scenario, is a brain-scientist system capable of giving rise to some kind of consciousness; we are not imagining that there could be a consciousness, identical to that which we enjoy, arising out of neural activity in a brain all by itself.

Third, when we produce an episode in consciousness by direct action on the brain, what we really do is modulate already existing states of consciousness. We affect the existing consciousness of a person whose brain we manipulate. By intervening in this way, we affect consciousness; we don't generate it out of nothing. At most, what we are entitled to conclude from considerations of this sort, then, is that action on the brain can produce changes in consciousness; we are not entitled to conclude that consciousness depends only on actions of the brain itself.

The Devil Is in the Details

The defender of the Foundation Argument insists that, in principle at least, the effects in consciousness normally brought about by the action of the world on the brain can be produced in the laboratory. The world, our interaction with it, drops out. We fix consciousness when we fix the states of the brain.

I suspect that one reason why this argument seems compelling is that we fail to think it through in all its details. Let us grant that it would be possible to control a subject's brain and therefore make it the case that this subject's brain undergoes the states and transitions that would take place in the brain of a normally embodied person engaged in normal perceptual interaction with the world. But let us further insist that we be more explicit about just what we would be imagining.

Remember that the state of my brain right now is dependent on all manner of conditions and processes. It depends no doubt on my metabolic and digestive states, and also on where I happen to be and what I happen to be doing. Food and beverage consumption, exercise, sleep cycles, emotional state—all these affect what my brain is doing. Moreover—and this is important—my brain is affected not only by what happens to me: by sights, sounds, odors, etc. What I do also brings about changes in my brain. For example, when I move my hands or eyes and head, I change my relation to the environment in which I find myself, and so I bring about changes in how things look and sound and smell. My brain state is not, in any straightforward sense, an effect brought about by patterns of stimulation along this or that input channel; it isn't determined by causal influence in one direction only. The character of my brain's condition is fixed by an ongoing dynamic of action and interaction between me and the environment around me, both the physical environment in which I find myself and also the biological environment of my bodily milieu.

This brings us to the crux: to act on the brain so as to simulate the effects of normal interaction with an environment would be tantamount to supplying, for the brain, an alternative bodily milieu and environment. Indeed, such an ersatz body and environment would be tantamount to a virtual world.

So, far from showing that we can make sense of the idea that consciousness arises out of the brain alone, we rely here on the idea that consciousness depends on the interplay among brain, body, and world, or at least among brain, body, and virtual world.

The consciousness we are imagining is the product of a complex dynamic interaction of the brain and our virtual ersatz environment. It would seem we've not taken any steps in the direction of thinking that a self-sufficient brain, entirely on its own, could suffice for consciousness. All we've shown is that perhaps there could be consciousness achieved by means other than the normal ones. Crucially, the consciousness that we are imagining is achieved by the ingenuity of neuroscientific engineers constructing an environment that is meant to affect us just as the regular environment would.

Real Minds, Virtual Reality

Our virtual-reality scenario falls short of demonstrating that there can be mind without the presence and involvement of the world for a deeper and more interesting reason. We imagine that by constructing a virtual reality we can induce the brain to be in the states it would be in if it was located, as it usually is, in a normal environment. But to imagine this is not to show that the brain can be so induced to have normal experiences. The virtual-reality scenario we have conjured is not one in which we see and feel and act just as we normally do; rather, it is a scenario in which it misleadingly seems to us as if that were the case. Virtual reality gives us, at best, virtual experience and virtual minds. And what that shows is that brain states alone are not sufficient for real experience and real minds. A trainee pilot in a flight simulator is not flying a real airplane, whatever he or she may think. And ditto for our more fanciful thought-experimental subject: instead of experiencing a world, we are imagining a way of being cut off from the world.

At this point, the defender of the consciousness-is-in-the-head thesis is likely to insist that I am not entitled to make this last claim. To do so is just to beg the question. That is, the defender will grant that what we have imagined is little more than virtual consciousness of a virtual world. But, the defender will insist, what reason is there to think that the experience of humans and animals is ever anything more than this sort of *Matrix*-like virtual simulacrum?

And so we hit rock bottom and our shovel goes no further. The true commitment that grounds the neuroscientific fantasy that we are, really, brains in vats is the familiar idea that the world we know is a fantasy, a grand illusion. In Chapter 6, I urged that nothing forces us to accept this conclusion, least of all any findings of contemporary perceptual psychology or cognitive neuroscience. It represents, it seems, something like an article of faith.

Dreaming

What about dreaming? Does the fact that we dream show that consciousness depends really only on what goes on inside us? This idea may strike many people as quite compelling. Each of us has had the experience of waking to find that what had seemed so vivid and so important and so real to us was only a dream. But in dreams we are not actively exploring the world or dynamically interacting with it. So it would seem that we must give up the idea that active exploration of the world, interaction with it, is necessary for consciousness.

Let us accept, for the sake of argument, that when we dream, we are utterly passive in relation to the world and that, for this reason, dreaming depends only on what is going on within us. From this premise it would not follow that where consciousness is concerned, all that matters is what is going on in the head. What would follow, at most, is that dream experiences depend only on what is going on in the head. The further claim—that all consciousness arises out of the action of the brain alone—does not follow unless we assume, as many traditional philosophers do, that any experience can occur in a dream. Does the neuroscientist's confidence that we can reduce mind to brain come down to this bit of traditional philosophy?

I reject the idea that all experience can occur in a dream. The traditional reason for supposing that any particular experience can happen in a dream is the fact—if it is a fact—that it is impossible to tell whether we are merely dreaming any given experience. This was Descartes' argument in his "First Meditation." What Descartes appreciated was that there was no way to tell whether one is really seeing or feeling or hearing what one thinks one does, or whether one is merely dreaming it. For any possible test that one performs—pinching oneself, for example—might also be a dream. And so, since we can't know on the basis of experience that we are not dreaming, Descartes concluded that we must look elsewhere, beyond the senses, to justify our beliefs about the world around us.

But even if we concede that it is impossible to tell by looking that one is not dreaming, that wouldn't show that there are not important differences between dream and nondream experiences. It would only show that it is difficult or impossible to tell. The point is a logical one. From the fact that I can't tell by looking whether a building is Georgian or Edwardian, it does not follow that there is no difference between Georgian and Edwardian buildings. But beyond this logical point, it is just simplistic to think that the actual phenomenon of dreaming supports this kind of skeptical conclusion.

Take another example: I have dreamt that I hear my mother calling me. The only way for me to describe what is happening in this dream is for me to characterize it as an episode of hearing my mother call me. But surely I can admit this without also admitting that, therefore, to hear my mother calling me in real life is to have an experience no different from the one I had in the dream. That's the sort of thing that generations of philosophers have felt compelled to say but that a little bit of reflection shows is entirely unjustified.

One striking difference between normal perceptual experience and experience in a dream is that the former, in contrast with the latter, is stable. Indeed, according to one dream researcher, Stephen LaBerge, it is a universal feature of dream experiences that detail is never stable across scenes in the dream. For example, if you read a sign in a dream and then, in the dream, turn away and then turn back, the words on the sign will have altered. This should not be surprising, from my perspective at least. After all, in normal perception—in contrast with dreams—we don't have to do the work of stabilizing detail. The detail is there, in the world. Reality anchors us. Whatever actions we take—shutting our eyes, turning away, getting distracted—things around us remain unaffected. In a dream, however, detail shows up as a feat of creative imagination. The fluidity and shifting grounds of dream experience reflect precisely the fact that in dreams, but not in normal perceptual experience, we are decoupled from the world around us. What determines content in a dream is precisely not what is there in front of us. We can see what we want to see, or what we are afraid to see, or what we wonder what it might be like to see. Which is just another way of saying that dream seeing is not really seeing at all.

In general, there is good reason to doubt that one can have bona fide perceptual experiences in a dream even if we feel compelled, when reporting on our dreams, to say that we do. In Chapter 6, I argued that we ought to think of perceiving as an activity of exploring the environment. It is not a process whereby a picture of the world is built up in your brain; rather, it is the activity whereby you achieve access to what is around you by making use of various different skills (of movement, of understanding, etc.). When I see a landscape, for example, I don't represent trees and meadow and clouds and sky and stream and birds and butterflies all at once in my head in the way these elements might all be depicted in a painting or drawing. They are present for me, in my visual experience, thanks to the fact that they are there and I have skillful access to them. The content of our experience—what we experience—is the world; in the world's absence, we are deprived of content. For this reason, whatever we think and feel and say, when we dream, we are not seeing.

In the absence of a demonstration that the class of perceptual experiences is a subset of the class of dream experiences, appealing to dreams shows only that what is going on in us is sufficient for dream experiences. And to learn that dreaming depends only on what is going on inside us is not yet to learn that all experiences depend only on what is going on inside us. If the truth be told, I am skeptical that the dreaming argument shows even that dreams depend on our internal states alone. Consider: there is probably reason to believe that what we can experience in a dream is limited by our past experience of the world. If that's true, then dreaming shows only that a certain narrow class of experience—dream experiences—can and must occur when an animal whose life is normally spent in close engagement with the world is for a time decoupled in sleep. Such considerations provide only pallid support for the idea that the brain is the minimal substrate of experience.

So the appeal to dreams, like the appeal to neuroscientific interventions, leaves us more or less where it starts: with unspecific Cartesian intuitions about the interiority of our experience.

Conclusion: Even the Mind of a Brain in a Vat
Needs a Body and a World

There is no empirical or philosophical justification for the idea that the brain alone is enough for consciousness. I hope I have convinced you that there is something perverse about the very idea that we are our brains, that the world we experience is within us. We don't need to have the world within us: we have access to the world around us; we are open to it. I take this to be the import of Bellow's language in this chapter's epigraph.

The idea that we are our brains is not something scientists have learned; it is rather a preconception that scientists have brought with them from home to their workbenches. It belongs not to well-established theory, nor even to that category of proposition—such as "I exist"—whose truth can require no verification. It is just prejudice. And in fact we have every reason to reject it now. It is a prejudice that constrains us like a straitjacket when we are trying to understand what we are and how we work. We spend all our lives embodied, environmentally situated, with others. We are not merely recipients of external influences, but are creatures built to receive influences that we ourselves enact; we are dynamically coupled with the world, not separate from it. In so many aspects of our lives this is becoming clear. Neuroscience must come to grips with it.

EPILOGUE: HOME SWEET HOME

We are out of our heads. We are in the world and of it. We are patterns of active engagement with fluid boundaries and changing components. We are distributed. So I have urged in this book.

If this book is an adventure story, then much of its drama unfolds in our attempts to escape the clutches of a durable yet false conception of our own intellectual predicament. Scientists seem to represent us as if we were strangers in a strange land. They represent us as if we were alienated. Nowhere is this view clearer than when discussion turns to relations among people. When we talk, for example, it is supposed that we transmit mere noises whose significance must be figured out by our conversation partner. According to this description of our interactions, we take up in relation to the other a stance of theoretical detachment and curiosity. Making sense of you is a puzzle I need to solve. Infants, coming to understand the world around them, are styled as scientists in the crib.

But we are not in this way alienated from each other or the world around us. We don't confront mere noises; we confront each other. We are always already in a shared context, and this shared context eliminates the need to figure out what is going on. Like a soccer player in mid-match, we are always already involved in the game. We rarely face the problem of assigning meaning to otherwise meaningless noises or deciding whether a being is conscious on the basis of the observation of behavior alone. We don't see mere patterns of shape and color and then judge them to be other people or objects. And so it is misguided to think that the brain's job is to solve these problems for us.

The substrate of our lives, and of our conscious experience, is the meaningful world in which we find ourselves. The broader world, and the character of our situation in it, is the raw material of a theory of conscious life. The brain plays a starring role in the story, to be sure. But the brain's job is not to "generate" consciousness. Consciousness isn't that kind of thing. It isn't a thing at all. The brain's job is to enable us to carry on as we do in relation to the world around us. Brain, body, and world—each plays a critical role in making us the kind of beings we are.

Our relation to the world is not that of an interpreter. The meaningful world is there for us, understood, before interpretation gets its start. The literary approach to the world—the world as a text standing in need of interpretation—is a dead end. Interestingly, many scientists working on problems about mind—cognition, thought, consciousness—presuppose something like the literary, interpretative stance to the world. But we don't secure the world through interpretation. Interpretation comes after we have the world in hand.

Our relation to the world is not that of a creator. The world is bigger than we are; what we are able to do is be open to it—that is, we are able to find our way around in it.

In mathematics you can distinguish the proof itself from the prose that surrounds the proof and comments on it. Philosophers writing about mathematics frequently take issue with the prose, but the proof itself stands untouched by philosophical scruples. In this book I am not interested in the prose of the science of consciousness but in the findings themselves. My purpose is not to comment on trends in neuroscience but to convince you that the neuroscientific, and more broadly the cognitive scientific, approach to mind needs rethinking from the ground up. Of course, there is much excellent experimental and theoretical work that has been and is now being done in cognitive science. But if I am right, whole research programs have to be set aside. It is misguided to search for neural correlates of consciousness—at least if these are understood, as they sometimes are, to be neural structures or processes that are alone sufficient for consciousness. There are no such neural structures. How could there be? It is a mistake to think that vision is a process in the brain whereby the brain builds up a representation of the world around us. It is likewise a mistake to think, as many neuroscientists now claim, that humans and other mammals are born with innate modules in the brain such as that for face detection. More generally, it is untenable to suppose that the brain's job is to do our thinking for us, and so it is untenable to think that the brain manages this task by performing complex computations.

As we move forward, then, we will appreciate that the foundations of consciousness are not distinctively neural. Insofar as we seek to understand the brain basis of experience, we will ask how the brain subserves our dynamic transactions with the world around us. We will keep the whole organism in focus and will think of the nervous system in the context of its normal embodiment. The developmental and evolutionary perspectives will be paramount, and we will pay close attention to the comparison of different species of animal. Just as we do not draw an impermeable boundary around the brain, we will not draw such a boundary around the individual organism itself. The environment of the organism will include not only the physical environment but also the habitat, including, sometimes, the cultural habitat of the organism.

Throughout this book I have been attacking orthodoxy and I

have been seeking to give shape to an alternative. I am not alone in doing this. The fabric of neuroscience and cognitive science is patchwork and variable. Even as orthodoxy spreads its branches, heterodoxy seeks to reach out of the shadow into the sunlight. The last twenty-five years have witnessed the gradual shaping of an embodied, situated approach to mind. This approach has flourished in certain regions of cognitive science such as philosophy and robotics, but it has been all but ignored in neuroscience, in mainstream linguistics, and, more generally, in the domain of consciousness studies. If we are to understand consciousness—the fact that we think and feel and that a world shows up for us—we need to turn our backs on the orthodox assumption that consciousness is something that happens inside us, like digestion. It is now clear, as it has not been before, that consciousness, like a work of improvisational music, is achieved in action, by us, thanks to our situation in and access to a world we know around us. We are in the world and of it. We are home sweet home.

NOTES

ACKNOWLEDGMENTS

INDEX

NOTES

PREFACE

Randy Nesse, of the University of Michigan Medical School, is now conducting illuminating research into evolutionary approaches to depression. He shows convincingly, I think, that neither medical nor other styles of biological reasoning provide support for the dictum that depression is a brain disease.

1: AN ASTONISHING HYPOTHESIS

The epigraph of this chapter is from Wittgenstein's *Philosophical Investigations* (New York: Macmillan, 1954, page 178).

Francis Crick and Christof Koch acknowledge how little we understand the neural basis of consciousness in "A framework for consciousness," in *Nature Neuroscience* 6, no. 2 (2003):119–126. They write: "No one has produced any plausible explanation as to how the experience of the redness of red could arise from the action of the brain" (page 119). Crick's discussion of the supposedly astonishing hypothesis is to be found in his book *The Astonishing Hypothesis: The Scientific Search for the Soul* (New York: Touchstone Press, 1994). Crick's dismissive statement about philosophy is to be found in "Visual perception: rivalry and consciousness," in *Nature* 379 (1996): 485–486. In this chapter I also cite Patricia Churchland's statement that evidence now supports the claim that it is the brain that does our thinking for us rather than something nonphysical. She makes this statement on the first page of *Brain-Wise: Studies in Neurophilosophy* (Cambridge, MA: MIT Press, 2002). This is an introduction to philosophy with an eye to the way knowledge of neuroscience can help us rethink traditional philosophical problems. If I am right, neuroscience today depends on a somewhat stagnant set of philosophical presuppositions.

Ned Block's distinction between *access* and *phenomenal* consciousness was drawn first in his 1994 article "On a confusion about a function of consciousness," in *Behavioral and Brain Sciences* 18, no. 2: 227–287. Thomas Nagel's famous essay on consciousness in which he characterizes consciousness in terms of "what it is like" is "What it is like to be a bat," in *The Philosophical Review* LXXXIII, no. 4 (October 1974): 435–450.

Wittgenstein's discussion of what looks and behaves like a human being is to be found in his *Philosophical Investigations*, section 281 (New York: Macmillan, 1954).

Descartes was a philosopher and scientist of enormous brilliance; his work repays careful reading and rereading. He is thought to be responsible for what is now known, after him, as Cartesian dualism. Dualism is the thesis that there are, in the universe, two fundamentally distinct kinds of stuff. There is mind and there is matter. How they relate is a problem; how we know about each is also a problem. Descartes' most important development of this idea is in the first and second of his *Meditations on First Philosophy*. This is available in many editions and translations.

We find it easy to ask about brains in vats—could it turn out that I am a brain in a vat? are we in fact brains in vats?—but we seldom press for details and ask ourselves just what kind of vat would be required to keep a brain going in anything like the way our brain is kept going. A lovely and original investigation of this by the philosopher Evan Thompson and the neuroscientist Diego Cosmelli is to be found in "Embodiment or envatment: reflections on the bodily basis of consciousness."

I rely in the text on Geraint Rees, Gabriel Kreiman, and Christof Koch's 2002 paper "Neural correlates of consciousness in humans," in *Nature Reviews Neuroscience* 3: 261–270. They suggest that visual consciousness can't be tied to narrow brain areas but that it seems rather to implicate large-scale and long-distance processes in the brain.

I rely in this chapter on papers of the Belgian neurologist Steven Laureys. I recommend, in particular, his "The locked-in syndrome: what it is like to be conscious but paralyzed and voiceless," found in *Progress in Brain Research* 150 (2005): 495–511. See also his "Brain function in the vegetative state," in *Acta neurologica belg.* 102 (2002): 177–185. Laureys was a member of a team that reported in a recent issue of the journal *Science* that some vegetative patients show appropriate brain activations when asked to imagine playing tennis; crucially these activations *cease* when the patient is told to *stop* imagining playing tennis. Different areas of the brain light up (the so-called parahippocampal place area) when the patient is instructed to think of being in her home. There is controversy surrounding these findings. What exactly do they show? This much is clear: they provide further grounds for taking seriously the obligation to look out for the needs of people in the persistent vegetative state. See A. M. Owen, M. R. Coleman, M. Boly, S. Laureys, J. D. Pickard, and M. H. Davis, "Detecting awareness in the vegetative state," in *Science* 313, no. 5792 (2006): 1502.

There are a number of memoirs of experience of life with locked-in syndrome. Most of these were dictated by the sufferers through elaborate eye-blinking codes. One of the most impressive of these works is Jean-Dominique Bauby's *The Diving Bell and the Butterfly* (New York: Vintage, 1997). This book has now been made into a successful movie directed by Julian Schnabel.

I have found Guy C. Van Orden and Kenneth R. Paap's critical paper on PET very helpful in preparing this chapter. See their "Functional neuroimages fail to discover pieces of mind in the parts of the brain," in *Philosophy of Science* 64 (1997): 85–94. I also recommend Robert Stufflebeam and William Bechtel's "PET: exploring the myth and the method," in *Philosophy of Science* 64 (1997): 95–106, and also Jim Bogen's "Epistemic custard pies from functional brain imaging," in *Philosophy of Science* 69 (2002): 59–71.

2: CONSCIOUS LIFE

The epigraph for this chapter is from Wittgenstein's *Philosophical Investigations*, page 178.

For resources on locked-in syndrome and persistent vegetative state, please see the references for Chapter 1.

Fritz Heider and Marianne Simmel's now classic paper "An experimental study of apparent behavior" was published in *American Journal of Psychology* 57 (1944): 243–259.

Rodney Brooks is a theoretical trailblazer in robotics and cognitive science. His papers make an effort to use robots as a tool for understanding human consciousness. Crucially, he makes it his objective to break with the traditional "intellectualist" assumption that, for a robot to be an intelligent agent, it must (as people are wrongly supposed to do) first sense, then construct a theory about what is going on, then plan, and then, finally, act in light of its aims or ends. Brooks builds robots that are maximally offloaded and distributed; that is, they are mobile and environment-specific, responding to local realities. If they exhibit intelligence—none of them actually do, and it is a worthwhile project to ask why—then they do so not thanks to something happening inside them but rather thanks to the way they are active and responsive to a particular situation. The intelligence would emerge from the dynamic interaction; it is a feature of the robot's active being and not, as it were, a by-product of something happening inside the robot.

One of Brooks's star students is Cynthia Breazeal, whose robot Kismet is mentioned in this chapter. For a nice essay on Kismet, see her "Robot in society: friend or appliance?" This is an MIT AI lab publication available for download on Breazeal's website. See also "How to build robots that make friends and influence people" by Breazeal and her colleague Brian Scassellati. This is also an MIT download.

"Theory of mind" is a hot topic not only in developmental psychology circles but also, now, in animal cognition research (cognitive ethology) and also in the area of autism research. The idea that our relation to each other is basically a theoretical one is an old idea in philosophy. Perhaps the best-known proposal to solve the problem of other minds is the so-called argument from analogy. The idea—developed in different ways by John Stuart Mill (see his *Systems of*

Logic) and Bertrand Russell (see his *Problems of Philosophy*)—is that we attribute mental attitudes to others on the basis of what they say and do. We are guided, in doing this, by the supposition that things are with others mentally as they would be with ourselves were we acting the way they are acting. As a pattern of inference, this analogical reasoning is pretty lousy. How could I possibly think that my own case—the correlations between mental states and behavior that I find in myself—is a guide to all other people? But the deeper problem—I try to develop this in the chapter—is that a view like this is mistaken in supposing that we need to reason or argue or infer in order to gain access to the minds of others.

In any case, in psychology, the theory of mind literature took off with the publication in 1980 of Heinz Wimmer and Josef Perner's investigations of belief in children. They developed the so-called false-belief task mentioned in the chapter. See H. Wimmer and J. Perner, "Beliefs about beliefs: representation and constraining function of wrong beliefs in young children's understanding of deception," in *Cognition* 13 (1983): 103–128. See also J. Perner, S. R. Leekam, and H. Wimmer, "The case for a conceptual deficit," in *British Journal of Developmental Psychology* 5 (1987): 125–137. For information on the issues as they arise in cognitive ethology, see Marc Hauser's "Our chimpanzee mind," in *Nature* 437 (2005), and Daniel John Povinelli's "Behind the ape's appearance: escaping anthropocentrism in the study of other minds," in *Daedalus* (Winter 2004): 29–41. The autism literature is burgeoning. For a reliable and insightful guide, I recommend S. Gallagher's "Understanding interpersonal problems in autism: interaction theory as an alternative to theory of mind," in *Philosophy, Psychiatry, and Psychology* 11, no. 3 (2004): 199–217.

Kenneth Kaye and Peter Hobson, discussed later in Chapter 3, provide resources for seeing how we need a superior, less theoretical way of conceptualizing our commitment to the minds of others. Very important, for this purpose, is Colwyn Trevarthen's influential "Communication and cooperation in early infancy. A description of primary intersubjectivity," an article published in M. Bullowa's *Before Speech: The Beginning of Human Communication* (Cambridge: Cambridge University Press, 1979, pages 321–347).

For information about the astonishing practice of trying animals in courts of law, see E. P. Evans's *The Criminal Prosecution and Capital Punishment of Animals*, originally published in 1906 in New York by Dutton.

I cite Vicki Hearne's excellent work on dogs in the chapter. I recommend, in particular, her *Adam's Task: Calling Animals by Name*. A fluent, philosophically insightful writer and a professional dog and horse trainer, she makes an important argument for the view that we enter into morally significant relationships with work and companion animals and that it is impossible to do justice to the tone and structure of these relationships by taking up a detached, behavior-oriented perspective. No one has written better about the lives of domestic animals than Hearne. Of course, what she says about animals goes all the more so for humans: the kinds of relationships we have with each other are incompatible, practically speaking, with a kind of detached, behavior-oriented, mechanistic standpoint.

My discussion of vervet monkeys relies on D. L. Cheney and R. M. Seyfarth's classic volume, *How Monkeys See the World* (Chicago: University of Chicago Press, 1990).

The central and ambitious theme of this chapter—that life is mind—has been developed in the work of others. I single out, in particular, the excellent and for me influential discussion in Evan Thompson's *Mind in Life* (Cambridge, MA: Harvard University Press, 2007). I would also mention a very good doctoral dissertation chapter by the Harvard philosophy student Bharath Vallabha (now a member of the philosophy faculty at Bryn Mawr).

3: THE DYNAMICS OF CONSCIOUSNESS

The epigraph for this chapter is from Emerson's essay "Experience."

This chapter grows out of work I did jointly with Susan Hurley between 2000 and the summer of 2007. See, in particular, our paper "Neural plasticity and consciousness," in *Biology and Philosophy* 18 (2003): 131–168, where a fair bit of the argument of this chapter is developed, and also "Synaesthesia and sensorimotor dynamics: how hunter-gatherers can hear color," in Michael Smith, Frank Jackson, and Robert Goodin (eds.), *Common Minds: Themes from the Philosophy of Philip Pettit* (Oxford: Oxford University Press, 2007). In this chapter I contrast two conditions, one in which the wiring up of a cortical area to a nonstandard source of stimulation causes the cortical area to change its qualitative function, and one in which the rewiring brings about no such qualitative change. Hurley and I, in our 2003 paper, called the first condition *cortical deference* and the second *cortical dominance*. Using this terminology, our leading question was: Why do we sometimes get deference and sometimes dominance? The general claim that I make in this chapter is that the *deferential brain* is the *healthy brain*; whenever we get successful adaptation and integration, we get deference.

I discuss the Nobel Prize–winning work of David Hubel and Torsten Wiesel in Chapter 7 and give references there.

In this chapter, and elsewhere in the book, I also build on work with Kevin O'Regan: see our "A sensorimotor account of vision and visual consciousness," in *Behavioral and Brain Sciences* 24, no. 5 (2001): 883–975. In my own *Action in Perception* (Cambridge, MA: MIT Press, 2004) I defend the general approach to visual perception that I present in this chapter.

Bruce E. Wexler's *Brain and Culture: Neurobiology, Ideology, and Social Science* (Cambridge, MA: MIT Press, 2006) presents a powerful and sustained argument for the position that we can understand the brain only in the setting of the animal's social and environmental life. I reviewed Wexler's book in the

Times Literary Supplement no. 5479 (2007): 24. Readers of this book may find not only Wexler's book but also my review of some interest. I learned of Kenneth Kaye's fascinating work on infant-mother dyads, breastfeeding, and, more generally, infant development from Wexler's book. Interested readers may want to read Kaye's *The Mental and Social Life of Babies: How Parents Create Persons* (Chicago: University of Chicago Press, 1982). Another excellent book on infant development that has influenced my own thinking is Peter Hobson's *The Cradle of Thought: Exploring the Origins of Thinking* (New York: Oxford University Press, 2004). In these books one finds ample support for my claim that the social/environmental context of the child's relationship with its caretaker is *necessary* for normal development; this is the basis for my claim that, really, we are not entitled to think of the child's mode of being as independent of this contextual embedding.

To learn more about Mriganka Sur's work on ferrets, see M. Sur, A. Angelucci and J. Sharma, "Rewiring cortex: the role of patterned activity in development and plasticity of neocortical circuits," in *Journal of Neurobiology* 41, no. 1 (1999): 33–43. This is a review article and contains pointers to other publications.

Sur's laboratory is not alone in conducting research along these lines. Of special interest is work from the laboratory of Leah Krubitzer at the University of California, Davis. In a series of studies with congenitally deaf mice, she found that as these mice matured, parts of the cortex that in normal mice would have become the auditory cortex took over both visual and sensorimotor function. As a result of disuse, it seems, the auditory cortex was called on to serve other sensory functions. See, for example, D. L. Hunt, E. N. Yamoah, and L. Krubitzer, "Multisensory plasticity in congenitally deaf mice: how are cortical areas functionally specified?" in *Neuroscience* 139, no. 4 (2006): 1507–1524.

Is perception with the tactile-visual substitution system a way of seeing? Can we maintain that the somatosensory cortex, although physiologically unchanged, takes on *visual* significance? In answer to the second question, I would predict that disruption of the somatosensory cortex (e.g., using TMS) in tactile-visual substitution ought to produce visual (or quasi-visual) disruption. To my knowledge, this has not been performed. As for the first question, there are clearly differences between this sort of tactile-visual and normal seeing, but there are also similarities. And moreover, whatever we say about the visual character of perceptual exploration in this technology-aided way, it is clearly not normal *tactile* perception. We need a principled way, then, to explain similarities and differences. This is what I offer here. Crucially, what does the explaining is never the intrinsic character of neural activity.

Paul Bach-y-Rita's original findings were published in *Nature* in 1969. For a general statement of his position, see his 1972 book, *Brain Mechanisms in Sensory Substitution* (New York: Academic Press). See also his "Tactile-vision substitution: past and future," in *International Journal of Neuroscience* 19, nos. 1–4 (1983): 29–36. There have been other efforts at devising sensory substitution systems. For an excellent recent overview by a philosopher and a scientist, I recommend Malika Auvray and Erik Myin's "Perception with compensatory devices: from sensory substitution to sensory extension." A striking finding of Bach-y-Rita's is that subjects were only able to see with the tactile-visual sensory substitution device if they wore the camera mounted to their heads or bodies. That is, the switch whereby mere tactile stimulation became a way of seeing crucially depends on the perceivers' bodily control of the tactile stimulation. This is exactly what a view such as mine would predict: what makes the experience visual is the way movements affect sensory stimulation and neural activation.

The phrase "the explanatory gap" is due to philosopher Joe Levine. See his "On leaving out what it's like" in the collection *Consciousness: Psychological and Philosophical Essays*, edited by Martin Davies and Glyn Humphreys (Oxford: Blackwell, 1993, pages 121–136).

4: WIDE MINDS

The epigraph for this chapter is taken from Dewey's essay "The new psychology." This was published in the *Andover Review* 2 (1884): 278–289. I made use of a copy of the text available online at the York University Classics in the History of Psychology website, at psychclassics.yorku.ca/.

The rubber-hand illusion was first reported by M. Botvinick and J. Cohen in *Nature* 391, no. 6669: 756. It has also been investigated by V. S. Ramachandran. His work on this topic, as well as on the topic of phantom limbs, is discussed in his book (cowritten with Sandra Blakeslee) *Phantoms in the Brain* (New York: William Morrow, 1998).

The McGurk Effect was first reported in 1978 by Harry McGurk and John MacDonald in "Hearing lips and seeing voices," in *Nature* 264: 746–748. Dominic Massaro at the University of California, Santa Cruz, has argued persuasively that *vision* plays a crucial role in normal speech perception. He puts this finding to use in his work with deaf children.

My own study of the problem of phantom limbs was carried out jointly with Susan Hurley. We discuss this in the 2003 paper I cited in the last chapter. Merleau-Ponty's ideas about phantom limbs and the body schema are to be found in his *Phenomenology of Perception*, especially in the chapter called "The body as object and mechanistic physiology." I quote in the text from page 82 of the translation by Colin Smith (London: Routledge and Kegan Paul, 1962). For an excellent development of Merleau-Ponty's ideas, and an introduction to him, see writing by Shaun Gallagher, especially his book *How the Body Shapes the Mind* (Oxford: Oxford University Press, 2005). The neuroscientist Marcel Kinsbourne has incorporated some of these ideas in his writing. I recommend, in particular, his "Awareness of one's own body: an attention theory of its nature, development and basis," in *The Body and The Self*, edited by J. L. Bermúdez, A. Marcel, and N. Eilan (Cambridge, MA: MIT Press, 1995, pages 225–245).

Gareth Evans observed that we do not need to think about how to get ourselves through doors in the way that we need to think about how to get a couch through a door. The latter, but not the former, is a problem for spatial cognition. See his *Varieties of Reference* (Oxford: Oxford University Press, 1982).

Dennis R. Proffitt and his University of Virginia PhD student Jessica K. Witt have shown that the perceived size of a baseball correlates with batting average. See their paper "See the ball, hit the ball" in the December 2005 issue of *Psychological Science*. They have also shown that inclined paths look steeper to tired hikers: see "Perceived slant: a dissociation between perception and action," in *Perception* 36, no. 2 (2007): 249–57. Yoshiaki Iwamura, Atsushi Iriki, and Michio Tanaka, in a 1996 paper in *Neuroreport*, showed that tool use can modify the body schema. For a general review of this area of research, see Nicholas Holmes and Charles Spence's "The body schema and the multisensory representation(s) of peripersonal space," in *Cognitive Process* 5, no. 2 (June 2004): 94-105.

Andy Clark and David J. Chalmers's essay "The extended mind," in *Analysis* 58, no. 1 (1998): 7–19, presents an argument that we should consider artifacts outside the head to belong to the "machinery of cognition" (to use Rick Grush's phrase). Anyone who appreciates that sometimes we think with words, or with our pens, or with our paintbrushes, can appreciate this insight. Clark has just pubished a new book on this "extended mind" hypothesis; it includes a foreword by Chalmers: see *Supersizing the Mind: Embodiment, Action, and Cognitive Extension* (New York: Oxford University Press, 2008). Notably, neither Clark nor Chalmers has sympathy for the idea developed here that consciousness itself can be explained only if we make use of such an extended conception of the machinery of mind. Conscious experience would seem to be detachable from and independent of the world without. I return to this topic in Chapter 8.

There's a lot of literature about new media and the ways they are changing our relations with others and our sense of ourselves. My comments in the text about instant messaging and Japanese teenagers draw on the research of Peter Lyman and Mimi Ito. I am grateful, in particular, to Peter Lyman (now deceased) for his helpful guidance.

Wendy Mackay's work on air traffic controllers is marvelous. I first learned of it from one of her collaborators, Anne-Laure Fayard. See, in particular, Mackay's "Is paper safer? The role of paper flight strips in air traffic control." This can be found in *ACM Transactions on Computer-Human Interaction* 6, no. 4, (1999): 311–340.

The locus classicus of Hilary Putnam's deservedly influential criticism of the classical conception of language is "The meaning of 'meaning,'" published in his *Philosophical Papers, Volume 2: Mind, Language and Reality* (Cambridge: Cambridge University Press, 1975). Wittgenstein's *Philosophical Investigations* is a sustained criticism of what I am here calling the classical conception of language. See, in particular, the first two hundred or so numbered paragraphs.

Eric Kandel's 2000 Nobel Prize lecture, "The Molecular Biology of Memory Storage: A Dialog between Genes and Synapses" (downloadable from the Nobel Prize website) lays out beautifully the work with snails for which he won the prize. There is a nice discussion of this in Bruce E. Wexler's book *Brain and Culture: Neurobiology, Ideology, and Social Change* (Cambridge, MA: MIT Press, 2006).

5: HABITS

The epigraph for this chapter is a stanza from the poem "Laying Down a Path in Walking" by Antonio Machado, from his *Proverbios y Cantares*, 1930. I discovered this poem reading Evan Thompson's *Mind and Life* (Cambridge, MA: Harvard University Press, 2007). The literal translation given here is that of the late Francisco Varela, taken from Varela's essay "Laying Down a Path in Walking," in William Irwin Thompson (ed.), *Gaia, A Way of Knowing: Political Implications of the New Biology* (Hudson, NY: Lindisfarne Press, 1987, pages 48–64). Varela, a neuroscientist of the first rank as well as a philosopher, has been an influence on my own thinking. The lines from the poem are used here with the permission of Amy Cohen Varela and William Irwin Thompson.

In philosophical writing, no one has done more to emphasize the importance of skill and habit and to criticize intellectualism than Hubert Dreyfus. Dreyfus's work in this area takes its start from his reading of Heidegger's *Being and Time* and Merleau-Ponty's *The Phenomenology of Perception*. For an introduction to Dreyfus's thinking, a good place to start is his *Being-in-the-World: A Commentary on Heidegger's* Being and Time, *Division I* (Cambridge, MA: MIT Press, 1991). My criticism of artificial intelligence in this and other chapters owes a debt to his seminal book *What Computers Can't Do* (Cambridge, MA: MIT Press, 1972), recently reissued as *What Computers Still Can't Do* (Cambridge, MA: MIT Press, 1992).

Differences in the focus of attention among experts and novices has been a theme in a good deal of psychological research. See, for example, Rob Gray's "Attending to the execution of a complex sensorimotor task: expertise differences, choking and slumps," in *Journal of Experimental Psychology: Applied* 10, no. 1 (2004): 42–54. Evidence that overall levels of neural activity decrease for expert performers can be found in "The mind of expert motor performance is cool and focused" by John Milton, Ana Solodkin, Petr Hlustik, and Steven L. Small, in *NeuroImage* 35 (2007): 804–813.

In this chapter I venture criticisms of a family of ideas about language that are almost sacrosanct in contemporary cognitive science and philosophy. For example, I criticize the idea that to understand a language is to be able to assign meanings to strings of words based on knowledge of the meanings of the individual words as well as the rules for their combination. I also criticize the idea that languages are abstract symbolic systems rather than aspects of specific, local, real-world activity. I rely in this chapter on ideas about language first developed by Roy Harris, for many years the professor of general linguistics at the University of Oxford. Harris is an original and important thinker whose work has not received the attention it deserves. I first encountered him as a student at Oxford in the late 1980s. His books *The Language Makers* (London: Duckworth, 1980) and *The Language Myth* (New York: St. Martin's, 1981) provide a fascinating criticism of what many linguists take for granted. For example, it is Harris who (to my knowledge) first pointed out that our idea that languages are intertranslatable is itself an artifact of the fact that we have established, in schools and elsewhere, important cultural practices of translation. In the absence of those practices, the correspondences between languages themselves are not, as it were, just given: they are made. Terrence Deacon's book on language, *The Symbolic Species* (New York: Norton, 1998), referred to in the text, is also of interest.

There is a huge empirical literature on object recognition, much of it focused on the question of whether neural mechanisms of face recognition are different from mechanisms involved in the perception of nonfacial objects. Nancy Kanwisher's work in this area is noteworthy. A place to start might be her paper "The fusiform face area: a module in human extrastriate cortex specialized for face perception" (cowritten with Josh McDermott and Marvin M. Chun) in *Journal of Neuroscience* 17, no. 11 (1997): 4302–4311). In this chapter I also cite from her short essay "What's in a face," in *Science* 311 (2006): 617–618 (I give a quotation from page 617), and from her "Neural events and perceptual awareness," in *Cognition* 79 (2001): 89–113 (my citation is from page 109). For support for the claim that newborns show a preference for faces, see C. C. Goren et al., "Visual following and pattern discrimination of face-like stimuli by newborn infants," in *Pediatrics* 56, no. 4 (1975): 544–549. On seeing faces upside down and the "inversion effect," see R. K. Yin, "Looking at upside-down faces," in *Journal of Experimental Psychology* 81, no. 1 (1969): 141–145, and also J. W. Tanaka and M. J. Farah, "Parts and wholes in face recognition," in *Quarterly Journal of Experimental Psychology* 46, no. 2 (1993): 225–245. For convincing criticism of Kanwisher's innate face-module approach, papers by Michael J. Tarr and his colleagues and students are to be recommended. See, in particular, "Learning to see faces and objects" by

Michael J. Tarr and Yi D. Cheng, in *Trends in Cognitive Sciences* 7, no. 1 (2003), and also "Beyond faces and modularity: the power of an expertise framework," in *Trends in Cognitive Sciences* 10, no. 4 (2006). The classic statement of the "expertise hypothesis"—the idea that our powers of face recognition are a special case of our more general ability to become expert in different kinds of object perception—is Susan Carey's "Becoming a face expert," in *Philosophical Transactions of the Royal Society of London* 335, no. 1273 (1992): 95–103. Specifically, in support of the claim that FFA is active when nonfacial objects are in a domain of expertise, see I. Gauthier et al., "Expertise for cars and birds recruits brain areas involved in face recognition," in *Nature Neuroscience* 3, no. 2 (2000): 191–197. See also J. W. Tanaka and T. Curran, "A neural basis for expert object recognition," in *Psychological Science* 12, no. 1 (2001): 43–47. For support for the claim that damage in FFA can produce specific deficits in perception for objects other than faces, see I. Gauthier et al., "Can face recognition really be dissociated from object recognition?" in *Journal of Cognitive Neuroscience* 11, no. 4 (1999) 349–370. For evidence that people find it more difficult to recognize different kinds of faces (i.e., faces from other "races"), see G. Rhodes et al., "Race sensitivity in face recognition: an effect of different encoding processes," in A. F. Bennett, K. M. McConkey, et al. (eds.), *Cognition in Individual and Social Contexts* (Amsterdam: Elsevier, 1989, pages 83–90).

For an excellent discussion of the existence of a "visual word form area" in the brain, see "The visual word form area: expertise for reading in the fusiform gyrus" by Bruce D. McCandliss, Laurent Cohen, and Stanislas Dehaene, in *Trends in Cognitive Sciences* 7, no. 7 (2003): 293–299.

The idea of cognitive trails has been explored by Adrian Cussins; indeed, the term, as I use it, is his. See his "Content, embodiment and objectivity: the theory of cognitive trails," in *Mind* 101, no. 404 (October 2002): 651–688. Evan Thompson, in his recent book *Mind in Life* (Cambridge, MA: Harvard University Press, 2007), makes use of a similar idea.

6: THE GRAND ILLUSION

The epigraph for this chapter is from Merleau-Ponty's *Phenomenology of Perception*, translated by Colin Smith, page 355.

I begin by citing a passage from Eric R. Kandel, James H. Schwartz, and Thomas M. Jessell's influential textbook *Essentials of Neural Science and Behavior* (Norwalk, CT: Appleton and Lang, 1995). Their words provide a nice statement of a view that I think is not empirically supportable. The first passage I quote comes from page 368, the second from page 321. A more recent and more philosophically delicate statement of just the same problematic idea that we are on the receiving end of an illusion promulgated by the brain is Chris

Frith's book *Making Up the Mind: How the Body Creates Our Mental World* (Oxford: Blackwell, 2007).

For an excellent scholarly introduction to the history of vision theory, see D. C. Lindberg's *Theories of Vision from Al-Kindi to Kepler* (Chicago: University of Chicago Press, 1976).

The use of the term "grand illusion" in this context is due to me and my colleagues Evan Thompson and Luiz Pessoa. For a general exploration of this issue, I recommend a collection of essays that I edited a few years ago for the *Journal of Consciousness Studies*. It brings together a number of philosophers and scientists to discuss this issue. See *Is the Visual World a Grand Illusion?* (Thorverton, Exeter, UK: Imprint Academic, 2002).

Change blindness was first discussed in print in a series of articles by Ron Rensink, Kevin O'Regan, and their colleagues. The seminal papers are J. K. O'Regan, R. A. Rensink, and J. J. Clark, "Mud splashes render picture changes invisible," in *Investigative Ophthalmology and Visual Sciences* 37 (1996): S213, and R. A. Rensink, J. K. O'Regan, and J. J. Clark, "To see or not to see: the need for attention to perceive changes in scenes," in *Psychological Science* 8, no. 5 (1997): 368–373. Of further interest is "A sensorimotor approach to vision and visual consciousness" by O'Regan and me, in *Behavioral and Brain Sciences* 24, no. 2 (2001): 939–1030. See also O'Regan's *Macmillan Encyclopedia of Cognitive Science* entry on "change blindness." Another good review of the literature in this area is D. J. Simons and R. A. Rensink, "Change blindness: past, present, future," in *Trends in Cognitive Sciences* 9, no. 1 (2005): 16–20.

Dan Simons is responsible for the most interesting work on "inattentional blindness." For the gorilla case, see in particular D. J. Simons and C. F. Chabris, "Gorillas in our midst: sustained inattentional blindness for dynamic events," in *Perception* 28 (1999): 1059–1074. See also Arien Mack and Irwin Rock's extended book-length treatment, *Inattentional Blindness* (Cambridge, MA: MIT Press, 1998).

In my text I try to make clear that there are two routes to the idea that the world is a grand illusion. The first is the more traditional route: we are given so much less than we think we see, so what we think we see must be something that arises in us thanks to the workings of the brain. This view is expressed by almost every major thinker working in this field (with a couple of notable exceptions, e.g., the psychologist James J. Gibson and the philosopher Maurice Merleau-Ponty). The second is the less traditional and indeed in some ways more radical idea that precisely because the brain is *not* in the business of building up pictures in the head, our experience really is profoundly illusory— that is, we don't even have the experience we think we do. Daniel Dennett is the person who has developed this idea most brilliantly. We (perceivers) think there is no gap in our experience, so we (scientists) suppose that the brain fills in the gap. But if it turns out that the brain does not fill in the gap, then we are left with the conclusion that we are radically deceived about what our experience is.

   In this chapter I argue that we can't convict ordinary perceivers of anything like this kind of error. First, ordinary perceivers don't think of themselves as representing the world in the head. Second, it is not the case, therefore, that our experience would be accurate only if, in fact, our brains filled in the gap in an internal representation of the world. For discussion of this and related issues, see Dennett's *Consciousness Explained* (Boston: Little, Brown, 1992). For criticism of Dennett, see my *Action in Perception* (Cambridge, MA: MIT Press, 2004). Kevin O'Regan has also advocated this particular, radical version of the "grand illusion" idea in his important paper "Solving the 'real' mysteries of visual perception: the world as an outside memory," in *Canadian Journal of Psychology* 46, no. 3 (1992): 461–488.

   For an interesting discussion of the homunculus fallacy and its influence on the theory of vision, see John Hyman's *The Imitation of Nature* (Oxford: Blackwell, 1989).

7: VOYAGES OF DISCOVERY

The epigraph for this chapter is from James J. Gibson's *Ecological Approach to Visual Perception* (Philadelphia: Lawrence Erlbaum, 1979, page 225).

   Hubel and Wiesel's papers have recently been collected by the authors and published under the title *Brain and Visual Perception: The Story of a 25-Year Collaboration* (New York: Oxford University Press, 2004). Included in the volume are interesting biographical essays as well as stage-setting forewords and afterwords to key papers (written mostly by Hubel). My discussion of Hubel and Wiesel in this chapter in fact started life as a review of this volume for the *Times Literary Supplement,* and I draw some paragraphs from that review here. I also draw on Hubel's *Eye, Brain, and Vision*, Scientific American Library, vol. 22 (New York: W. H. Freeman, 1988). Hubel and Wiesel's writing belongs to the canon in vision science, and brief reformulations of their basic findings are included in all textbooks in this field. I found Steven E. Palmer's treatment especially helpful and I recommend it. See Palmer's *Vision Science: From Photons to Phenomenology* (Cambridge, MA: MIT Press, 1999). David Marr's book *Vision* (San Francisco: W. H. Freeman, 1982) is insightful, rigorous, and accessible.

   Most of what I write here about the history of neuroscience I learned from the distinguished neuroscientist Walter Freeman. I am grateful to him for generous and instructive conversation.

   My discussion of computers and artificial intelligence in the text is indebted both to John R. Searle's famous criticism of AI—see, for example, *The Rediscovery of Mind* (Cambridge, MA: MIT Press, 1992)—and also to Daniel

Dennett's brilliant defense of it: see, for example, *Brainstorms* (Cambridge, MA: MIT Press, 1978). I agree with Searle that doing what computers do can't be what makes us conscious. But I agree with Dennett that it is an open question whether computers or robots can one day become conscious. Even if they can, we know, a priori and thanks to Searle, that we need to look elsewhere than to their computational innards for an understanding of that in which their *mindedness* consists. Searle seems to think that when it comes to us, we need look no further than the brain for an understanding of the ground of our consciousness. But that's a mistake—one Dennett warns us against—and it reveals a mistaken assumption in his criticism of the possibility of computational minds. Information processing in the brain does not a mind make, but that's because *nothing in the brain* makes the mind. The great insight of AI is that we are, in a way, on a par with machines. If a robot had a mind, it would not be thanks to what is taking place inside it alone (thought of computationally or otherwise). It would be thanks to its dynamic relation to the world around it. But that's exactly the case for us as well. For an excellent and accessible introduction to this general topic, including a nice introduction to the relevant basic mathematics, see Jack Copeland's *Artificial Intelligence: A Philosophical Introduction* (Oxford: Blackwell, 1993).

For a brilliant review of *Blade Runner* with an eye to its philosophical upshots, see Stephen Mulhall's *On Film* (2nd ed.) (Abingdon, UK: Routledge, 2008). *Blade Runner*, directed by Ridley Scott, was based on a book by Philip K. Dick called *Do Androids Dream of Electric Sheep?*

8: A NOTHING RESERVED FOR EVERYTHING

The epigraph for this chapter is from Saul Bellow's novel *Humboldt's Gift*. I haven't read the novel but I found the passage typed out by hand among my grandmother Marion Hageman's papers after she died. It is an extraordinary passage and I often wonder what my grandmother had in mind when she copied it down and saved it. The quote appears on page 313 of Penguin Books' 1996 edition.

Francis Crick made the claim with which this chapter begins in "Visual perception: rivalry and consciousness," in *Nature* 379 (1996): 485–486.

The Foundation Argument, as I call it, is the line of thought that underlies a good deal of thinking about consciousness and the brain. It can fairly be called Cartesian, for it was Descartes, in the "Second Meditation," who enacted the line of thought that each of us has a kind of immediate awareness of himself, an awareness that depends on the truth of no other beliefs about our physical nature or even of the very existence of the world around us. The famous statement of Descartes *Cogito, ergo sum*—I think or feel or experience or undergo sensations, therefore I am—seems to license the idea that our fundamental nature is our consciousness and that all the rest, our situation, our bodies, the place of others in our lives, is just accident. If you accept this Cartesian conception that all I really know is my own experience—that all the world is real only insofar as I experience it within my mind—then it is just a step to the idea that the brain is alone necessary for experience. For if we assume that consciousness has some physical basis, then it would seem we are safe in assuming that its physical basis is independent of all that is extrinsic to consciousness.

There are in fact several philosophers—foremost among them Andy Clark—who grant that the body and the external environment play an important role in constituting our cognitive apparatus. But experience itself—pure consciousness, Clark would have it—depends only on factors inside us. It is Clark, in personal correspondence, who has sharply articulated what I am here calling the Foundation Argument—i.e., the idea that the fact that we dream and that we can produce experiences by direct action on the brain shows that consciousness depends only on what is happening in the brain.

Stephen LaBerge, a psychologist trained at Stanford University, is a dream researcher. His focus has been lucid dreaming. LaBerge informed me in conversation that his research shows that a sign never reads the same way twice in a dream.

A few of the sentences in this chapter also appear in an essay I have written called "Magical realism and the limits of intelligibility: what makes us conscious," published in *Philosophical Perspectives* 21: *Philosophy of Mind* (2007).

ACKNOWLEDGMENTS

This book is dedicated to the memory of Susan L. Hurley, who died on August 16, 2007, at the age of fifty-two in Oxford, England. Susan was my friend and my teacher. Philosophy and science are poorer now that she is gone. As I indicate at different junctures in the text, this book reflects what I learned during the course of my collaboration and friendship with her.

Miriam Dym, my wife, has been my closest friend and companion and I would never have written this book without her.

Some readers of this book will see the influence of others in it. None is stronger than that of Hubert Dreyfus and Evan Thompson. I also feel a debt of gratitude to Ned Block, Daniel Dennett, Kevin O'Regan, and John Searle. They show up all over this book, sometimes as allies and sometimes as targets of criticism. I have been nourished intellectually by my friends and students at the University of California, Berkeley, in particular (in addition to Hubert Dreyfus and John Searle) John Campbell, Walter Freeman, James Genone, Kristina Gerhman, Farid Masrour, John Schwenkler, and James Stazicker. My thinking about the topic of this book also benefited from my engagement with UC Berkeley's Center for New Media and my participation in working-group discussions with its members, as well as by many conversations with Dan Zahavi at the Center for Subjectivity Research in Copenhagen.

Judith Baldwin Noë and Hans Noë, my parents, and Alexander Nagel, my friend, have offered me excellent criticism. Thanks also to Gwen Shupe for her thoughtful response to the book.

Thanks to John Brockman and Russell Weinberger for support and encouragement. And thanks, in particular, to Joe Wisnovsky, my editor at Hill and Wang, for his commitment to this project.

The bulk of this book was written while I was a fellow at the Wissenschaftskolleg zu Berlin (the Institute for Advanced Study in Berlin). I am deeply indebted to the rector, fellowship, and staff of this unusual and fine institution. And thanks to the University of California, Berkeley, and the Department of Philosophy in particular, for stimulation and support, as well as for making it possible for me to accept the hospitality of the Wissenschaftskolleg.

INDEX

M

Machado, Antonio, 97
Macintosh operating system, 70, 110
Mackay, Wendy, 85–86
magnetic stimulation, transcranial, 173, 194n
Making Up the Mind (Frith), 130
mammals, 185; immaturity at birth of, 94; plasticity of brain in, 49–50; visual system of, research on, 131, 149–57
Man with Two Brains, The (movie), 10, 12
marine snails, 91–94
marmosets, 116
Marr, David, 157, 160, 166
Martin, Steve, 10–12
Massachusetts Institute of Technology (MIT), 27, 53, 151
mathematical theory of information, 156
mathematics, 88, 184
McCulloch, Warren, 156
McGurk effect, 73
meaning, 52, 164, 183–84; linguistic, 89–90, 102, 108, 126
Merleau-Ponty, Maurice, 75, 76, 82, 129, 142, 200n
mice, congenitally deaf, 194n
Middle Ages, 131, 132
Mill, John Stuart, 191n
mind: attribution of, 27–28; computer model of, 157–59, 161, 164–65; concept of brain versus, 9–10; extension of, 87–89; as life, 41–45; of others, moral commitment to, 32–35; paradox of science and, 39–41; "theory" of, 29–30, 191n; and virtual reality, 176–77

mind-body problem, 165–66
monetary value, practices and conventions as basis of, 34
monkeys, 116; research on vision in, 151, 154; social interaction of, 43–46; tool use by, 79
motion, illusions of, 173
Mountcastle, Vernon, 151

N

Nagel, Thomas, 9
natural selection, 40
navigating, 81–82
Nazi Germany, 38, 68
Newton, Isaac, xv, 131
New York University, 92
Niger, 115; languages spoken in, 103
Nigeria, 115
Nobel Prize in Physiology and Medicine, 5, 131, 149, 156, 157, 167, 169
novelty, radical, 124
novices, *see* performance, of novices versus experts
numerical notation, 87–88, 158

O

object recognition, 198n
optic nerve, 134
O'Regan, Kevin, 60
Oxford University, 198n